Flow, Deformation and Fracture

Over 40 years of teaching experience are distilled into this text. The guiding principle is the wide use of the concept of intermediate asymptotics, which enables the natural introduction of the modeling of real bodies by continua.

Beginning with a detailed explanation of the continuum approximation for the mathematical modeling of the motion and equilibrium of real bodies, the author continues with a general survey of the necessary methods and tools for analyzing models. Next, specific idealized approximations are presented, including ideal incompressible fluids, elastic bodies and Newtonian viscous fluids. The author not only presents general concepts, but also devotes chapters to examining significant problems, including turbulence, wave propagation, defects and cracks, fatigue and fracture. Each of these applications reveals essential information about the particular approximation.

The author's tried and tested approach reveals insights that will be valued by every teacher and student of mechanics.

G. I. BARENBLATT is Emeritus G. I. Taylor Professor of Fluid Mechanics at the University of Cambridge, Emeritus Professor at the University of California, Berkeley, and Principal Scientist in the Institute of Oceanology of the Russian Academy of Sciences, Moscow.

Cambridge Texts in Applied Mathematics

All titles listed below can be obtained from good booksellers or from Cambridge University Press. For a complete series listing, visit www.cambridge.org/mathematics.

Nonlinear Dispersive Waves
MARK J. ABLOWITZ

Complex Variables: Introduction and Applications (2nd Edition)
MARK J. ABLOWITZ & ATHANASSIOS S. FOKAS

Scaling
G. I. BARENBLATT

Hydrodynamic Instabilities
FRANÇOIS CHARRU

The Mathematics of Signal Processing
STEVEN B. DAMELIN & WILLARD MILLER, JR.

A First Course in Continuum Mechanics
OSCAR GONZALEZ & ANDREW M. STUART

A Physical Introduction to Suspension Dynamics
ÉLISABETH GUAZZELLI & JEFFREY F. MORRIS

Applied Solid Mechanics
PETER HOWELL, GREGORY KOZYREFF & JOHN OCKENDON

Practical Applied Mathematics: Modelling, Analysis, Approximation
SAM HOWISON

A First Course in the Numerical Analysis of Differential Equations (2nd Edition)
ARIEH ISERLES

Iterative Methods in Combinatorial Optimization
LAP-CHI LAU, R. RAVI & MOHIT SINGH

A First Course in Combinatorial Optimization
JON LEE

An Introduction to Computational Stochastic PDEs
GABRIEL J. LORD, CATHERINE E. POWELL & TONY SHARDLOW

An Introduction to Parallel and Vector Scientific Computation
RONALD W. SHONKWILER & LEW LEFTON

Flow, Deformation and Fracture

Lectures on Fluid Mechanics and the Mechanics of Deformable Solids for Mathematicians and Physicists

GRIGORY ISAAKOVICH BARENBLATT, ForMemRS

Emeritus G. I. Taylor Professor of Fluid Mechanics,
University of Cambridge
Emeritus Professor, University of California, Berkeley
Principal Scientist, Institute of Oceanology,
Russian Academy of Sciences

CAMBRIDGE
UNIVERSITY PRESS

CAMBRIDGE
UNIVERSITY PRESS

University Printing House, Cambridge CB2 8BS, United Kingdom

One Liberty Plaza, 20th Floor, New York, NY 10006, USA

477 Williamstown Road, Port Melbourne, VIC 3207, Australia

314-321, 3rd Floor, Plot 3, Splendor Forum, Jasola District Centre, New Delhi - 110025, India

79 Anson Road, #06-04/06, Singapore 079906

Cambridge University Press is part of the University of Cambridge.

It furthers the University's mission by disseminating knowledge in the pursuit of education, learning and research at the highest international levels of excellence.

www.cambridge.org
Information on this title: www.cambridge.org/9780521715386

First published 2014

A catalogue record for this publication is available from the British Library

Library of Congress Cataloging in Publication data
Barenblatt, G. I.
Flow, deformation and fracture : lectures on fluid mechanics and the mechanics of deformable solids for mathematicians and physicists / Grigory Isaakovich Barenblatt, for. mem. RS, emeritus professor, University of California, Berkeley, emeritus G.I. Taylor Professor of Fluid Mechanics, University of Cambridge, principal scientist, Institute of Oceanology, Russian Academy of Sciences.
pages   cm. – (Cambridge texts in applied mathematics)
Includes bibliographical references and index.
ISBN 978-0-521-88752-6 (hardback) – ISBN 978-0-521-71538-6 (pbk.)   1. Fluid mechanics – Textbooks.
2. Deformations (Mechanics) – Textbooks.   3. Fracture mechanics – Textbooks.   I. Title.
QA901.B37   2014
532′.05 – dc23   2013048955

ISBN 978-0-521-88752-6 Hardback
ISBN 978-0-521-71538-6 Paperback

To the glowing memory of my grandfather,

Professor at Moscow State University,

Veniamin Fedorovich Kagan

this work is dedicated

with eternal love and gratitude

Contents

Foreword

In his preface to this book, Professor G. I. Barenblatt recounts the saga of the course of mechanics of continua on which the book is based. This saga originated at the Moscow State University under the aegis of the renowned Rector I. G. Petrovsky and moved with the author first to the Moscow Institute for Physics and Technology, then to Cambridge University in England, then to Stanford University, until it reached its final home as a much loved and appreciated course at the mathematics department of the University of California, Berkeley. Those not fortunate enough to have been able to attend the course now have the opportunity to see what has made it so special.

The present book is a masterful exposition of fluid and solid mechanics, informed by the ideas of scaling and intermediate asymptotics, a methodology and point of view of which Professor Barenblatt is one of the originators. Most physical theories are intermediate, in the sense that they describe the behavior of physical systems on spatial and temporal scales intermediate between much smaller scales and much larger scales; for example, the Navier–Stokes equations describe fluid motion on spatial scales larger than molecular scales but not so large that relativity must be taken into account and on time scales larger than the time scale of molecular collisions but not so large that the vessel that contains the fluid collapses through aging. An awareness of the scales that are relevant to each problem must guide the formulation of mathematical models as well as the asymptotics that lead to their solution. Accordingly, the book makes explicit the intermediate asymptotic nature of many of the well-known arguments in mechanics, leading to a clear understanding of their domains of validity and their limitations. Along the way, the assumptions that underlie the various models are spelled out in detail and expressed in terms of the appropriate dimensionless numbers. Dimensional and scaling considerations are introduced early, allowing the reader an easy way to understand the consequences of the various assumptions without the heavy mathematical machinery that can impede understanding if it is introduced prematurely.

These unique features make it possible for this book to present, without long preliminaries, not only the basic features of the mechanics of continua but also more sophisticated topics in turbulence and fracture, fields to which Professor Barenblatt has contributed so much but where present understanding is still incomplete. In particular, the reader will surely appreciate the illuminating presentations of crack propagation and of the scaling of turbulent shear flow. I was privileged to work on this last topic with Professors Barenblatt and Prostokishin, and I feel very strongly that our straightforward scaling arguments satisfy all logical requirements and fit all the data ("all" means literally "all") better than the older alternatives. The presentation here is detailed enough for the reader to form his or her own opinion.

Finally, I would like to mention that the historical asides that Professor Barenblatt provides throughout are enjoyable and illuminating. This is indeed a remarkable book.

Alexandre J. Chorin
University Professor
University of California

Preface

(1) Mechanics, the science of the motion and equilibrium of real bodies, is the oldest natural science created by humankind. Unlike other scientific disciplines, mechanics had no predecessors. Prehistoric human beings started to take their first steps in mechanics – apparently even before starting to speak – when they invented and improved their first primitive tools.

Later, when humankind took its first steps in mathematics, mechanics was the first field of application. The value of mechanics for mathematics was clearly emphasized by Leonardo da Vinci: "Mechanics is the paradise for mathematical sciences because through it one comes to the fruits of mathematics."

Nowadays mechanics is an organic part of applied mathematics – the art of designing mathematical models of phenomena in nature, society and engineering.

The development of mechanics, and the recognition of its value, has not gone smoothly over the centuries. An analogy of mechanics with the phoenix comes to mind. This legendary bird has appeared with practically identical magical features in the ancient legends of many cultures: Egyptian, Chinese, Hebrew, Greek, Roman, native North American, Russian and others. The names were different: the name "phoenix" was coined by the Assyrians; Russians called it "zhar-ptitsa" (fire-bird). According to the legend, unlike all other living beings the phoenix had no parents, and death could never touch it. However, from time to time, when it was weakened, the phoenix would carefully prepare a fire from aromatic herbs collected from throughout the world and burn itself. Everything superfluous is burnt in the fire and a new beautiful creative life opens to the phoenix.

So, what is the analogy? Mechanics now is living through a critical period. The community at large, particularly the scientific community, often considers mechanics to be a subject of secondary value. As a practical result of this attitude, bright young people nowadays are not interested in choosing mechanics as their profession. Inevitably, and rather quickly, a decline in the level of students leads to a decline in the level of professors. One after another mechanics departments in universities and technological institutes close; mechanics disappears from the curricula of physics and mathematics.

I am sure that a phoenix-type rebirth of mechanics is unavoidable. The reason for my confidence is the existence (not always generally recognized by society, nor its political leaders) of fundamental problems of vital importance for humankind that cannot be solved without the leading participation of mechanics and applied mathematics as a whole. To mention a few of these: the suppression of tropical hurricanes; the prediction of earthquakes three to four hours before the event; the creation of a new branch of engineering based on nano-technology; the creation of a new standard for the development of the deposits of the Earth's non-renewable resources, in particular oil, gas and coal, replacing the existing predatory exploitation; etc. These problems should be the subject of national and multinational programs. The participation of mechanics, renewed when necessary, in solving such problems is unavoidable: the strength and long-lastingness of mechanics come from its ability for continual renewal and, once renewed, for tackling potential problems of primary importance for leading nations and for humankind as a whole. The surge of interest in mechanics during World War II and its aftermath demonstrated this clearly.

(2) Ivan Georgievich Petrovsky, a great mathematician whose rectorship guided an epoch for Moscow State University, was a man of unsurpassable vision in science and education. His role as a leading mathematician is illustrated by a remarkable event: Sir Michael Atiyah, R. Bott and L. Gårding had published an article (Atiyah, Bott and Gårding, 1970) with a dedication to I. G. Petrovsky in Russian, clarifying and generalizing – as the authors said – Petrovsky's theory of lacunas for hyperbolic differential operators. His textbooks on ordinary differential equations (Petrovsky, 1966), partial differential equations (Petrovsky, 1967) and integral equations (Petrovsky, 1967) have seen wide use throughout the world. Petrovsky died in the building of the Central Committee of the Communist Party (he himself was not a member of the Party), fighting for his principles regarding the selection, teaching and education of students.

Petrovsky had a clear view of the role of mechanics in scientific education. It was he who offered the late V. I. Arnold, at that time a brilliant young professor of mathematics who became later one of the world's leading mathematicians, a chance to deliver the course of classical rational mechanics for student mathematicians. The mechanics community at Moscow State University considered this offer to be a risky experiment, but the result exceeded all expectations: Arnold's book (Arnold, 1978) *Mathematical Methods of Classical Mechanics* became a gem of scientific literature.[1]

---

[1] The title of the book was also a result of compromise.

Later, Petrovsky, as Chairman of the Department of Differential Equations, gave me the chance to teach the course on the mechanics of continua. The class was excellent: it contained very strong students, now disseminated over the entire planet. Ivan Georgievich offered me remarkable conditions: a one-year course, two exams, complete freedom in selecting the curriculum. However, he did impose one strict condition: "I cannot force the students to attend your lectures, so at each lecture you should tell them something special that they cannot find in the textbooks".

After the end of the Petrovsky era at Moscow State University I continued, through the initiative of my former student V. B. Librovich (at that time the leading person in the Academy's Institute for Problems in Mechanics), to deliver this course under the same conditions at the Moscow Institute for Physics and Technology (MIPT). This institute was a remarkable school, founded after World War II by an enthusiastic group of first-class physicists, applied mathematicians and engineers, headed by S. A. Christianovich. They were inspired by the example of the Ecole Polytechnique, and the level of the entering students was very high.

When I moved to the West, I delivered appropriate parts of this course in Cambridge, UK, the University of Illinois, Urbana–Champaign, Stanford University, and finally at the University of California, Berkeley. Naturally, each time the content of the course was suitably modified, but the general style that had been blessed by I. G. Petrovsky remained the same.

(3) Now I want to mention some specific features of the present book. Contrary to common practice, I do not devote a substantial initial part of the book to the presentation of tensor calculus. I prefer to introduce tensors and discuss their properties at the spot where they naturally appear.

At the very beginning of this book I introduce the concept of *intermediate asymptotics* and use it widely throughout. It is worthwhile to present here the definition of this concept, formally introduced by Ya. B. Zeldovich and myself, but *de facto* used long before that. Let us assume that in a problem under consideration there are two parameters X_1 and X_2 having the dimension of an independent variable x and that X_1 is much less than X_2, so that there exists an interval of values of x where $X_1 \ll x$ and at the same time $x \ll X_2$. The asymptotic representation, or simply asymptotics, in this interval is called the *intermediate asymptotics*.

It is emphasized that the fundamental concept of a *continuous medium* introduced after that, and widely discussed and illustrated by many examples, is one of intermediate asymptotics.

Also, at the very beginning, the concept of an *observer* is formally introduced, as well as the *invariance principle*: all physically significant relations can be written in a form valid for all observers.

(4) Next comes a detailed presentation of *dimensional analysis and physical simili-tude*, as a consequence of the invariance principle: physical relations can be written in a form valid for *all observers, having units of measurement of the same physical nature but of different magnitudes.* Dimensional analysis is widely used through-out the whole book, and it greatly simplifies the presentation. It is difficult for me to understand the custom of presenting these concepts at the end of mechanics of continua courses, where they can be used for nothing.

After introducing the fundamental concept of a continuous medium which, I emphasize, is based entirely on the intermediate asymptotic approach, multiple illustrations are presented justifying this concept on various scales from the atomic to the cosmic. After that the basic equations of mass and momentum con-servation for continuous media are presented. Here the concept of the mass flux vector and the stress tensor appear naturally, following from the invariance principle.

(5) To obtain a closed system of basic equations for mathematical models, the continuous medium should be supplied with some physical features. A general ap-proach using *finite* constitutive equations is outlined. It is emphasized that in fact this approach is of restricted value; it is justified only when the microstructure of the material remains intact in the process of the motion. Otherwise, the finite con-stitutive equations should be replaced by equations for the kinetics of microstruc-tural transformation. This comment should be taken into account, in particular, when one is analyzing the rheology of complex fluids, principally polymeric solu-tions and melts. In particular, the universality of the constants entering constitutive equations such as the Oldroyd B model should be verified for selected materials in a given class of motions; otherwise, predictions based on the model could be incorrect.

(6) After these preliminary steps, basic models or idealizations are presented. It is emphasized that all these models are *approximations*, valid for a restricted class of materials in a restricted range of motions and loading situations.

After this we follow with the classic models (approximations): the model of an ideal incompressible fluid, the model of an ideal elastic solid, the model of a New-tonian viscous fluid, and the model of an ideal gas. The conditions of applicability of each model are discussed in detail. In their due place the conservation laws of energy and angular momentum are presented and used in constructing the models. These conservation laws appear exactly where they are needed and can be used.

(7) The presentation of each model follows the same general scheme: general re-lations are derived, with specially selected "accompanying problems", presented in detail with a discussion of why they can be used. So, for the model of an ideal incompressible fluid, the accompanying problems are the Lavrentiev problem of a

directed explosion, and the problem of the lift force on a slender, weakly inclined wing. It is demonstrated, in particular, that the model of an ideal incompressible fluid needs to be modified in a natural way in the formulation of the lift force problem, which is substantially nonlinear in spite of the linearity of the basic equations. Here the concept of intermediate asymptotics is crucially important. The problem of brittle and quasi-brittle fracture accompanies the presentation of the ideal elastic solid approximation. It is demonstrated that the model should also be thus modified for the formulation of this substantially nonlinear problem, in spite of the linearity of the basic equations. (The situation is similar to the lift force problem.) It is demonstrated that the fracture of a structure is not a local event but a global one. From a mathematical viewpoint, fracture is the loss of existence of the solution to an explicitly formulated nonlinear free boundary problem for the linear equations of the equilibrium of an elastic body.

The problem of the boundary layer and drag of a slender weakly inclined wing accompanies the Newtonian viscous fluid approximation. The Prandtl–Blasius intermediate asymptotic solution to the problem of the boundary layer on a flat plate is presented in detail; the underwater reefs to be found in this problem are specially emphasized and discussed.

A more detailed analysis and classification of scaling laws is needed before we can proceed further. This is presented in detail in Chapter 9. The concepts of complete and incomplete similarity are introduced and discussed. They are widely used in the subsequent chapters, in which we discuss the ideal gas approximation and turbulence. The accompanying problem of impulsive loading is presented in Chapter 10 concerning the ideal gas approximation.

The final two chapters concern turbulence. I have found that the only two accompanying problems preserving the same style as in previous chapters that can be presented are those of turbulent shear flows at very large Reynolds numbers and of the local structure of turbulent flows at very large Reynolds numbers (the Kolmogorov–Obukhov theory). The discussion of turbulent shear flows at very large Reynolds numbers is based on the works of A. J. Chorin, V. M. Prostokishin and the present author. Some colleagues consider our model to be controversial. That is their business; the formulae and the comparison with experimental data speak for themselves.

So, this is the content of the present book as it was delivered in my courses of lectures.

Now it is time for acknowledgments. I want to remember with deep gratitude and admiration the late Ivan Georgievich Petrovsky who generated this course at Moscow State University. The support and care of the late Vadim Bronislavovich Librovich allowed me to continue this course at the Moscow Institute for Physics and Technology. I remember with gratitude Sergey Sergeevich Voyt who extended my audience at MIPT to the students affiliated with the Institute of Oceanology. I was honored by election to the G. I. Taylor Chair in Fluid Mechanics at the University of Cambridge: my cordial gratitude goes to G. K. Batchelor, D. G. Crighton, H. K. Moffatt and Sir James Lighthill. I spent a term at the University of Illinois, Urbana–Champaign: my deep thanks go to H. Aref and N. D. Goldenfeld. Later I was elected S. P. Timoshenko Visiting Professor of Applied Mechanics at Stanford University: my deep gratitude goes to M. D. Van Dyke, J. B. Keller and T. J. R. Hughes.

Before my period as Timoshenko Professor, I came for a short one-month Miller Visiting Professorship at Berkeley, by invitation of Professor Alexandre Chorin. We met for the first time on 16 February 1996; we both celebrate this date. In our first discussion we understood that our interests in turbulence studies and our general views on turbulence and science in general, although developed independently, practically coincide. We began to work together. Our first paper was submitted to the *Proceedings of the US National Academy of Sciences* three weeks later. Twenty papers followed this one and now, though working in different fields, we continue to systematically exchange thoughts. My gratitude to Alexandre is immeasurable: together with A. N. Kolmogorov and Ya. B. Zeldovich he became a benchmark in my scientific life. In particular I appreciate his many-fold advice concerning this book. Alexandre's efforts to obtain for me an honorary position at the University of California, Berkeley were strongly supported by Professor Calvin Moore, former Vice-President of the University of California, and at that time Chairman of the Department of Mathematics. Though himself a pure mathematician, Calvin Moore always strongly supported applied mathematics. I appreciate his friendship, kind attention and help rendered to me throughout my time at Berkeley.

I express my gratitude to Professor James Sethian, head of the Department of Mathematics at the Lawrence Berkeley National Laboratory. We work in different fields of mathematics, but his friendly attention to my work has been a permanent stimulus for me.

My cordial thanks go to Mrs Valerie Heartlie, Administrative Assistant of the department. She did her best to make this department a pleasant place for my work and everyday life.

I want to thank two people who played an essential role in my work on this book. These are Professor Valery Mikhailovich Prostokishin, who attended my lectures and gave me most valuable advice, and Dr David Tranah, Publishing Director, Mathematical Sciences and Information Technology, Cambridge University Press, who initiated the book, escorted it from its first version, and influenced its structure and style.

The help of Mrs Jean McKenzie, the Head Librarian of Berkeley Engineering Library, is highly appreciated.

I grew up in the family of my maternal grandfather, Veniamin Fedorovich Kagan (in the western mathematical literature he is known as Benjamin Kagan), after my mother, a physician–virologist, perished preparing the first vaccine for Japanese encephalitis. My grandfather was an outstanding mathematician, working in the foundations of geometry and non-Euclidean geometry; at Moscow State University he created an influential school of tensor differential geometry. He was a person of unbending principles in science and life and achieved an extraordinary moral authority, which allowed him to survive the stormy 1920s and 1930s of the Soviet era, even being accused of "sabotage in science" by the leading journal of the Communist party (*Bolshevik*) and thereafter immediately arrested.

The dedication of this book to his glowing memory is but a weak expression of my love and eternal gratitude to him.

Introduction

The purpose of the present book is to give an idea about fundamental concepts and methods, as well as instructive special results, of a unified *intermediate asymptotic* mathematical theory of the flow, deformation and fracture of real fluids and deformable solids. This theory is based on a quite definite and, we emphasize, idealized approach where the real materials are replaced by a continuous medium; therefore it is often called the mechanics of continua. It generalizes, and represents from a unified viewpoint, more focused disciplines: fluid dynamics, gas dynamics, the theory of elasticity, the theory of plasticity etc. For various reasons these disciplines underwent separate development for a long time. The splendid exceptions found in the work of A. L. Cauchy, C. L. M. H. Navier and A. Barré de Saint-Venant in the nineteenth century and L. Prandtl, Th. von Kármán, and G. I. Taylor in the twentieth century confirm rather than disprove the general rule. Therefore the teaching of the mechanics of continua and, more generally, the maintaining of interest in the mechanics of continua as a unified scientific discipline was, in this period of fragmentation, the job of physicists, who considered it to be a necessary part of a complete course of theoretical physics. So it was not by accident that among those who created courses in mechanics of continua were outstanding physicists: M. Planck, A. Sommerfeld, V. A. Fock, Ya. I. Frenkel, L. D. Landau and E. M. Lifshitz and, more recently, L. M. Brekhovskikh.

The physicists were more interested, however, in general ideas and methods rather than in the consistent presentation of special, even very important, results, which in fact give true shape to the subject.

In recent decades the tendency to take the separate branches of the mechanics of continua and combine them into a unified scientific discipline has penetrated to the mechanics community, and the practitioners of various specialities and scientific profiles have been involved in this activity.

There have been several reasons for this tendency; first of all, the internal need for the development of the mechanics of continua in order for one to understand

and evaluate, from a unified viewpoint, the methods and results achieved in the various specialized branches of this subject. Bearing in mind the modern grand scale of activity in mechanics stimulated by practical applications, it was simply not expedient to develop repeatedly from scratch essentially the same ideas and overcome similar difficulties in applications to different branches of the mechanics of continua.

An instructive example: the investigation of the slow motion of viscous fluids in wedge-shaped vessels and the investigation of the stress–strain state in elastic wedges can be reduced to completely identical mathematical problems. Nevertheless, for more than 60 years the study of these two problems was performed separately, without even a single cross-reference, in spite of the now obvious fact that instead of solving from scratch a problem in fluid dynamics it is enough to re-interpret appropriately the solution to the corresponding elasticity problem obtained long before, and vice versa.

It seems therefore expedient to continue to work to unify the subject as much as possible and in particular to avoid the appearance and propagation of separate terminology in its special branches – these constitute a language barrier which prevents the understanding of problem formulation and solutions by even qualified specialists not belonging to a certain narrow community. Such a unification should aid the discovery and use of results about analogous activities in other branches of the mechanics of continua.

Long ago Lord Rayleigh in Great Britain and L. I. Mandelstam in the Soviet Union developed a remarkable idea of "oscillation mutual aid" – a stimulating exchange of ideas and methods between physicists studying oscillations of different types: mechanical, acoustic, electromagnetic etc. The realization of this idea is an instructive example of a fruitful approach unifying various branches of physics on the basis of a unified method. Another instructive example of such a unified approach is synergetics.

The second, and perhaps the most important, reason for promoting the above-mentioned consolidation has been the growth in the range of the materials used in technology and also in the range of parameters (temperature, pressure etc.) of the working conditions of even classical structural materials such as steel. Thus it has become necessary to study the behavior of structures fabricated from synthetic materials and from more traditional materials under extreme conditions: high temperatures, pressures, strain rates and loading rates etc. For such applications the classical models of materials used in the theory of elasticity, fluid dynamics etc. are insufficient and more general ones are needed.

In more detail, the situation is as follows. Every material at fixed external conditions (temperature, pressure etc.) has a certain characteristic "relaxation time" τ. This is the time during which a deformed body in its fixed shape (Figure 1)

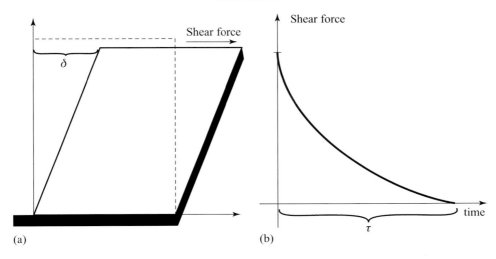

Figure 1 (a) A shear deformation imposed on a body, and the resulting fixed skewed shape of the body. (b) The shear stress disappears after a "relaxation time" τ.

continues to exhibit shear stresses before settling down to a new equilibrium. Every physical process of loading and/or deformation also has its own characteristic time T, e.g. the time duration of application of the load. Therefore the general behavior of a given material in a given loading or deformation process is determined by the ratio of these two times, the dimensionless parameter

$$De = \frac{\tau}{T}, \tag{1}$$

the *Deborah number*, as named by M. Reiner.[1] If $De \ll 1$ then the behavior of the material can be qualified as fluid and if $De \gg 1$ the material can be considered as a solid. So, for water under normal conditions we have $\tau \sim 10^{-12}$ s and for steel (in the range of tensile or compressing stresses of the usual order, 1000 kgf/cm^2) $\tau \sim 10^{12}$ s. The duration T of a human experiment is usually in the range 10^{-9} s (1 nanosecond) $< T < 10^9$ s (30 years). Therefore, for any T in this range, De for water is very small, less than one thousandth, and for steel it is very large, more than one thousand.[2]

---

[1] The reason for such a name derives from the Song of Deborah in the Old Testament (Judges **5**:5): "The mountains flowed from before the Lord". It is interesting that in the old King James version it was written "melted" instead of "flowed"; careful comparison with the original showed that this was not quite correct (especially for the purpose under discussion) and a proper translation was done in the later Cambridge academic editions of the Bible. The same mistake occurs in the Slavic Church text of the Old Testament.

[2] Now it becomes clear what exactly is the meaning of the statement of Deborah in modern scientific terms: as is known, the characteristic time for rocks (of which "mountains" consist) is of the order of 10^{11} s. However, the characteristic time of the Lord as mentioned by Deborah should be taken, according to modern estimates of the age of the Universe (the time since the Creation (the Big Bang)) to be several billion years, $T > 10^{16}$ s. Therefore, De is less than 10^{-5} and, indeed, rocks ("mountains") can be considered as fluids. This argument

At the same time, for Silly Putty, a well-known material from which a popular toy used to be made, τ is of the order of 1 s. Therefore this material can have both fluid and solid body behavior in an easily accessible range of observation times. Indeed, if it is thrown quickly to the floor it will rebound like a rubber ball; put it on a horizontal solid surface and it will flow slowly along the surface; hammer it sharply and it will break like fragile glass. Similarly, to observe the "solid body" behavior of water, for instance its brittle fracture, and the formation and propagation of cracks, it is necessary to load it with an extremely high strain rate: 10^{14}–$10^{15}\,\mathrm{s}^{-1}$. (Such experiments have been performed!)

For the familiar material polymethylmetacrylate (PMMA), an organic glass, the formation and propagation of cracks can be observed by simply using strain rates of the order of $10^4\,\mathrm{s}^{-1}$, easily accessible with the help of a laser pulse in the regime of free generation. However, for structural materials of a common steel type, at high temperatures or at high stresses, the relaxation time decreases rapidly and then, even at ordinary strain rates and loading rates, steel cannot be considered as a solid.

These reasons have led to an increasing interest in studying and teaching the flow, deformation and fracture of bodies as a unified subject, based on sufficiently general assumptions concerning the material under consideration, and in developing corresponding general approaches.

Scientifically speaking, mechanics is a synthesis of the approaches of physics, applied mathematics[3] and engineering. This is also completely true for the mechanics of continua. However, we will concentrate basically on the applied mathematics and physics view, using engineering ideas only to motivate the formulation and description of problems, and will not address detailed considerations of special engineering problems.

As mentioned above, our plan is to explore the fundamental ideas of the mechanics of continua, to give an idea of its basic methods and approaches as well as its instructive special results. Therefore we will not consider those topics that require for their exposition long calculations, with no compensation provided through the frequent use of the results achieved: such topics should be the subject of special courses and books.

obviously also demonstrates that Deborah was aware of correct estimates of the age of the Universe and, in particular, of the Earth; previous Biblical estimates (several thousand years, $\sim 10^{11}$ s) as well as the estimates of Lord Kelvin (hundreds of thousands of years), which did not take into account the heat generated by radioactivity, would be in conflict with her statement.

[3] There is nowadays considerable discussion concerning the subject of applied mathematics. In fact, its proper understanding is clarified if we remember the famous saying of J. W. Gibbs: "Mathematics is also a language." If so, then on the one hand pure mathematicians can be identified with linguists, who study the laws of the formation, history and evolution of languages. On the other hand applied mathematicians, building models of phenomena, are analogous to writers and poets who use language to create novels, poetry, critical essays etc. Applied mathematics is the art of constructing mathematical models of phenomena in nature, engineering and society.

The subject of fluid mechanics and the mechanics of deformable solids is a classical one. Some of the results we present were obtained many years, even many decades ago. It is well known, however, that each era brings its own perception to classical masterpieces of art: literature, music, paintings relate to what excites the human spirit today. This is equally true for a book on the mechanics of continua: even classical results are understood today quite differently from previous times and often need a completely different interpretation.

It is natural to ask what tools are required in continuum mechanics for modeling the equilibrium, motions, deformation, flow and fracture of real bodies.

First we mention the *general invariance principle*, whose value goes far beyond the borders of mechanics. The invariance principle is formulated in the following way:

> *The laws of equilibrium and motion can be expressed through equations valid for all observers.*

The invariance principle establishes the equivalence, the "possession of equal rights" by all observers, which is physically natural. Its formulation is trivial. We will see, however, that the consequences of this principle can be highly non-trivial.

Furthermore, we have at our disposal *conservation laws*: fundamental laws expressing the conservation (or balance) of mass, momentum, angular momentum, energy etc. These conservation laws give the most important fundamental relations of the mechanics of continua. They are, however, insufficient for obtaining a closed mathematical formulation of a model.

Therefore a very important aspect of any study in the mechanics of continua is to supply the continuous medium under consideration with physical properties. In the classical mechanics of continua this is done through the introduction of *finite constitutive equations*. Such equations are a priori relations that are approximately valid for a certain class of motions of a certain class of materials in which the researcher is interested. They connect the above-mentioned properties of state and/or the motion of real bodies with the internal forces acting in them. As classical examples, we can consider Hooke's law for an ideal elastic body, which establishes the proportionality of stress and strain, or Newton's law for viscous flow, which establishes the proportionality of the stress and the strain rate; see Figure 2. If these equations are applicable and well checked, they allow us in principle to obtain a closed system of equations sufficient to construct a mathematical model. It is tacitly assumed that the parameters entering the constitutive equations, such as the coefficients of proportionality in Hooke's law or in Newton's law mentioned before, are universal constants, valid for the whole class of phenomena under consideration. In fact it means that in these phenomena the microstructure of the material remains the same. There are wide and important classes of

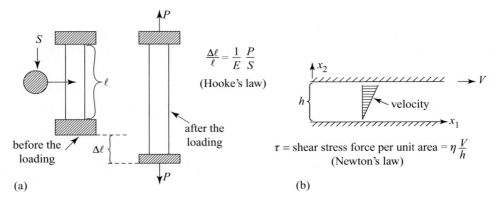

$$\frac{\Delta\ell}{\ell} = \frac{1}{E}\frac{P}{S}$$

(Hooke's law)

$$\tau = \text{shear stress force per unit area} = \eta\frac{V}{h}$$

(Newton's law)

(a)   (b)

Figure 2 (a) Hooke's law establishes the proportionality of the strain (the elon-
gation per unit length) of a rod and the stress (the tensile force per unit cross-
sectional area). (b) Newton's law establishes the proportionality of the strain rate
(the velocity difference per unit layer thickness) and the shear force per unit area.
The basic content of both laws is that the constants of proportionality η and E are
material properties, constant for fixed external conditions.

problems where this is so, and the models provided by the constitutive equations
are successful.

The important thing is, however, that in general this is not the case and that
therefore the kinetic equations governing the microstructural transformations of
the material should also be included in the mathematical model. A classical example is the Zeldovich–Frank–Kamenetsky model of gas combustion, where the
laws of mass conservation, energy conservation and the kinetics of the material
transformation are considered simultaneously.

Mathematical analysis plays a most important role in the mechanics of continua.
In constructing mathematical models of phenomena researchers have created a language of basic concepts that allows our intuition to be shaped. A good mathematical model starts to live on its own, revealing new features of a phenomenon that
are very often unexpected for the researcher who created it.

Proving *well-posedness*, i.e. the existence and uniqueness of the solution to a
closed system of equations of a mathematical model in a physically reasonable
class of functions, under appropriate initial and boundary conditions, is an essential stage of mathematical modeling. However, the problems of the existence and
uniqueness of the solution in the mechanics of continua play a much more important role than is generally recognized. There are problems where the really interesting answer is not the solution itself but the range of parameters where solutions
exist. Fracture and thermal explosions are classic examples of such problems.

Nevertheless, the central part of the mathematical analysis of problems in the mechanics of continua is not concerned with these topics. An applied mathematician working in this field is interested first of all in obtaining qualitative laws for the phenomena under investigation, i.e. the qualitative properties of the solutions. In a more or less general formulation such qualitative analysis is very difficult although it sometimes happens to be possible. The best chance for investigating the qualitative properties of a phenomenon is given by an explicit analytic form of the solution. To obtain a solution in such a form it is usually necessary to *simplify the problem* formulation as much as possible, in particular to simplify its geometry. Only the most basic features of the phenomenon should be left in the model, those without which the essence of the phenomenon would be lost.[4] It is a very important and responsible stage of modeling. As Einstein is alleged to have said "Things should be as simple as possible, *but no simpler!*" The value of solutions obtained in such a way very often goes far beyond the framework of the corresponding special problems because they can allow clarification of the properties of solutions of wide classes of problems. Textbooks on fluid mechanics, elasticity etc. are not collections of solutions to special problems, as pure mathematicians sometimes claim. Using analytical, and therefore the most transparent, solutions, applied mathematicians working in the mechanics of continua reveal general ideas. Very often special solutions happen to be the intermediate asymptotics to solutions of wide classes of problems where the influence of the non-essential details of the initial and boundary conditions has already disappeared but the system is still far from its state of ultimate equilibrium. Therefore the asymptotic approach pervades continuum mechanics.

For applications it is very important to *compute the solutions* to special problems with sufficient accuracy. Therefore computational methods in mechanics play a most important role. They should not be seen as in opposition to analytical methods. In fact a numerical investigation is especially fruitful after the qualitative features of the expected solutions have been clarified using analytical methods. Numerical computation in general has many features in common with experimental investigation; numerical schemes and algorithms should be carefully investigated and approved in the same way as good experimentalists calibrate their measuring devices. However, the analogy between a numerical computation and an experimental investigation is in fact much deeper than this. The researcher performing a numerical computation cannot be satisfied by a listing of the results of

---

[4] Compare the poem by Alexander Blok, "Retribution":

Rule out the accidental features
And you will see: the world is marvelous

(Translation from Russian by Sir James Lighthill.)

Figure 3 At some intermediate distance from the picture everyone will recognize the Mona Lisa. Up close the image disappears – it turns out to consist of
560 monochromatic squares distributed in a particular way. However, at large
distances the image naturally disappears again. This is an example of an "intermediate asymptotic" consideration, which will be basic in our presentation. From
Harmon (1973). Reproduced with permission. Copyright 1973 Scientific American, Inc. All rights reserved.

computations. These results should be processed and attempts made to extract from
them some general laws.

   A particular example is worthwhile noting: specialists in color printing assign
numbers to all colors, for even the finest shades of difference. Therefore the "exact"
listing of colors in Figure 3 will constitute a table of the following form:

Square number	Row number	Column number	Number for color of the square
1	1	1	2040–G20Y
2	1	2	4050–G20Y
⋮	⋮	⋮	⋮
560	28	20	2040–G20Y

It is clear that there is more present than just this listing: the Mona Lisa still needs to appear! We will return to this important example later.

The creation of a mathematical model requires the acceptance of certain assumptions, i.e. basic hypotheses. Such an acceptance is always a deliberate act. It should be justified by observation or experiment – a physical or numerical experiment based on a more detailed model. Therefore in the mechanics of continua an appropriate *combination of mathematical modeling with experiment*, be it physical or numerical, is decisive.

The reader may be astonished that the number of tools that have been mentioned is so small. But compare it with the number of tools available to artists: an easel, brushes, palette, canvas, paints – what else . . .? However, the history of civilization would be much poorer without the visual art which has been created using these tools, and only them!

Observation, mathematical modeling and experiment combine to form the basis of continuum mechanics. The goal of this book is to demonstrate how this is done.

1

Idealized continuous media: the basic concepts

1.1 The idealized model of a continuous medium

In the mechanics of continua the most important invention (whose fundamental value, however, is not always appreciated because it seems so natural) is the very concept of a continuous modeling of real materials. More precisely, the truly fundamental discovery was recognizing that to a knowable degree of accuracy the motion, deformation, fracture and/or equilibrium of *real bodies* can be based on an idealization (a *model*), that of a *continuous medium.*

In fact, we intend to study, i.e. to make models of, the motions, deformations, flows, fracture etc. of real bodies. These bodies consist of specific materials: honey, milk, petroleum, metals, polymers, ceramics, rocks, composites etc. If we look at these materials with the naked eye they very often seem continuous and homogeneous. But, when viewed through a microscope or telescope these materials (see Figures 1.1–1.7) display a developed microstructure at various scales – from atomic to essentially macroscopic ones – having a huge diversity of shape. How can we account for this diversity of shapes and properties of the elements of microstructures? Let us forget for the moment that we do not know the equations governing the equilibrium or motion. We do know, however, that taking into account the shape of the elements of a microstructure should mean accepting certain conditions at the boundaries of these odd formations. Let us imagine that by some miracle we know all these odd shapes. It is easy to show that it is impossible to write down the conditions at the boundaries of the microstructural elements even for the simplest problems.

Indeed, consider, for example, the problem of oil flow in an oil deposit. A large oil deposit has a volume of the order of a billion cubic meters. The oil moves in pores whose diameter is of the order of one to ten micrometers. The relative volume occupied by the pores is of the order of one tenth. Therefore in the deposit under consideration there are 10^{23} pores. Let us assume that for a description of a single

Figure 1.1 A microphotograph of the tip of a tungsten needle. (× 5 000 000). The intersections of crystal planes are clearly seen. J. J. Hren and B. A. Newman. Reproduced from Cottrell (1967), p. 92 with permission. Copyright 1967 Scientific American, Inc. All rights reserved.

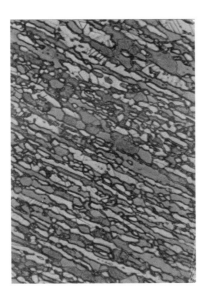

Figure 1.2 A microphotograph of a specimen of superplastic steel (× 3000). The ferrite (colored) and austenite (white) parts can be seen. Reproduced from H. W. Hayden, R. C. Gibson and J. H. Brophy (1969), p. 31 with permission. Copyright 1969 Scientific American, Inc. All rights reserved.

Figure 1.3 A microphotograph of a silicon polymer in polarized light. Neighboring crystalline and amorphous regions can be seen. F. P. Price. Reproduced from Mark (1967), p. 148 with permission. Copyright 1967 Scientific American, Inc. All rights reserved.

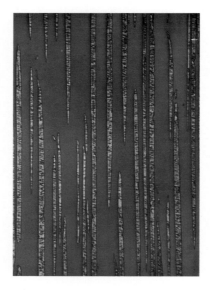

Figure 1.4 A microphotograph of a metallocomposite ($\times 2000$). High strength niobium "whiskers" are seen in a niobium matrix. P. Lemkey and M. J. Salkind. Reproduced from Kelly (1967), p. 160 with permission. Copyright 1967 Scientific American, Inc. All rights reserved.

Figure 1.5 A microphotograph of the ceramic uranium carbide, a combination of metallic uranium and carbon (× 300). W. E. Bruce. Reproduced from Gilman (1967), p. 112 with permission. Copyright 1967 Scientific American, Inc. All rights reserved.

Figure 1.6 A microphotograph of a semi-conductor, cadmium sulphide (× 800). The defects of crystallic structure of material are seen. C. E. Bleil, H. W. Sturner. Reproduced from Mott (1967), p. 81 with permission. Copyright 1967 Scientific American, Inc. All rights reserved.

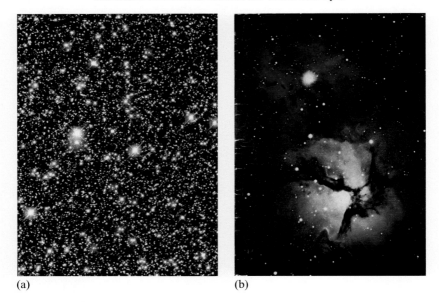

(a)   (b)

Figure 1.7 (a) The biggest ball of stars in our galaxy, the globular cluster *Omega Centaurus*, which contains about 10 million separate stars as it orbits the center of the Milky Way. From Trefil (1999), p. 221. (b) The Triffid Nebula in the plane of the Milky Way. Color photograph from the Palomar Hale telescope. From Bok (1972), p. 48.

pore only one printed character is needed; then the description of the deposit will require 10^{23} printed characters. As is well known, the average book contains 10^6 characters, so 10^{17} books of average size would be needed to write down these boundary conditions! All the paper in the world would not be enough. And even in the digital era, the figure of 10^{23} bytes exceeds the memory of the largest computers by many orders of magnitude!

An analogous situation occurs for the motion of gas in space. Indeed, even an ordinary thimble under normal pressure contains a huge number of gas molecules. The motion of each molecule is governed by Newton's second law, i.e. by an ordinary differential equation of the second order. To write down the initial conditions, i.e. the initial positions and velocities of each molecule (even if by a miracle we knew them), we would need more paper than there is on Earth, and that for a thimble of gas!

However, the very diversity and disorder of a real body's actual microstructure helps the researcher. Because of this disorder, the details of microstructural configuration can be assumed to be random (in fact this is a very strong assumption, which, generally speaking, requires confirmation). After making this assumption we can replace the real matter in the whole diversity of microstructure by an imaginary, fictitious "continuous medium" filling the space continuously!

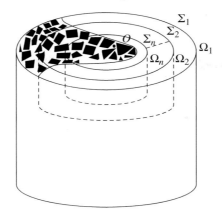

Figure 1.8 Introduction of a fictitious porous medium filling space continuously.

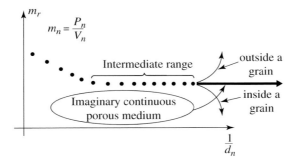

Figure 1.9 Porosity versus inverse diameter of the region Ω_n; see Figure 1.8.

An easy way to understand the introduction of the concept of a continuous medium is to consider a slow fluid flow, e.g. of oil, in a porous medium: so-called seepage or filtration. Indeed, let us take (see Figure 1.8) a certain geometric point O of a porous medium and consider a sequence of volumes Ω_n shrinking to it. For each volume Ω_n we can define the *mean porosity*

$$m_n = \frac{P_n}{V_n}, \tag{1.1}$$

where P_n is the volume of the part of region Ω_n occupied by pores and V_n is the total volume of Ω_n. It is in fact convenient to plot the dependence of m_n on the inverse diameter (i.e. the inverse maximal length size) d_n^{-1} of the volume Ω_n (see Figure 1.9).

When the volume of Ω_n is still large, the mean porosity depends on the size of the region because of the natural inhomogeneity of the rock. On the other hand, when the size of the region Ω_n becomes sufficiently small, less than the size of a single pore, the quantity m_n becomes equal to 0 or 1 depending on where the selected point is – inside a pore or inside a grain. It so happens, however, that for real porous media forming oil or gas deposits or aquifers in the middle of this ladder of scales, i.e. for Ω_n sufficiently small but not too small, "representative samplings" of a porous medium are contained: the quantity m_n appears to be practically independent of the size of the domain Ω_n and so equal to a constant m. It is worthwhile therefore to introduce into our considerations an imaginary "continuous" porous medium that fills the space continuously and has a *constant porosity m*. This means that inside an arbitrarily small region filled by this continuous medium there will be both pores and solid parts, so that the relative volume of the pores will be equal to m.

In an analogous way the concept of a kinematic characteristic of fluid motion in a porous medium – the *velocity of filtration* – can be introduced. Indeed, let us imagine a plane section of the porous medium. Take a geometric point O in this section and consider a sequence of areas Σ_n (the top faces of the volumes Ω_n) shrinking to this point (Figure 1.8). Let us divide the fluid volume flowing through an area Σ_n per unit time by the *whole area* Σ_n. At small, but not too small, sizes of Σ_n the obtained quantity for real porous media happens to be independent of the area size Σ_n. The reason is the randomness of the microstructure has already been accounted for, via the process of representative sampling, whereas the macro-inhomogeneity of the medium is not yet apparent. The quantity thus obtained is defined as the projection of the vector of the velocity of filtration normal to the area at the point under consideration. The projections onto three mutually perpendicular directions determine the filtration velocity vector.[1]

It is natural that, in introducing a simplification – an asymptotic concept of a continuous porous medium and the fluid flowing in it – and describing real motion on the basis of this concept, we lose a certain amount of information: we are *consciously* rejecting any attempt to describe the properties of a real medium and the fluid motion at a "microstructural" length scale, in this case the length scale of a single pore. *The application of the continuous medium approach is possible only if the information we lose is irrelevant to our purposes.*

Let us mention, however, that in fact the idea of a continuous medium is very natural. We essentially use the same approach in our reception of visual art, e.g. in painting. Our brain regards as a "continuous" picture the image produced by the artist's brush on the canvas. Of course, as we come near to the painting we do not expect to get an accurate view at the length scale of a brush stroke or smaller. Exceptions, like Jan Vermeer's painting "A girl reading a letter", where the text of the letter can be seen through a magnifying glass, are now considered rather amusing and this fact, at least for modern connoisseurs, does not increase the painting's

---

[1] We will see why the filtration velocity is a vector a little later: here I remind the reader that a vector is not simply a set of three numbers but also a law of transformation of these numbers when we pass to a different system of coordinates.

Figure 1.10 The brush stroke in Vincent Van Gogh's painting "Starry night" covers space scales of many light years.

value. An opposite example: a brush stroke in Vincent Van Gogh's painting "Starry night" (Figure 1.10) covers a space scale of many light years.

Now, look again at the painting shown in Figure 3 in the Introduction. This "painting" consists of 560 squares, each monochromatic. There is no accurate view at the length scale of a single square. However, at a distance sufficiently large, but not too large, everyone will recognize Leonardo's "Mona Lisa"!

The use of a continuum approach for modeling certain phenomena is then a sharply expressed act of will. Whether it is possible in any particular case is a problem that can be solved by comparison with experiment or a more detailed theory or, at least, on the basis of the researcher's intuition. It is obvious that to apply this continuum approach it is necessary that the size of the body under consideration should be much larger than the linear scale of the microstructure. For instance, it can be applied in modeling the motion of oil, water or gas in natural collectors. However, it cannot be used to describe water flow in a small pile of large cobblestones. The length scale at which the continuum approach ceases to be adequate can vary considerably. For instance, for air motion in the lower layers of the atmosphere this length scale is of the order of 10^{-5} cm; for fluid motion in porous rocks it is of the order of 10^{-3} cm; for modeling the strength of a concrete dam it is of the order of 1 cm. The continuum approach is used nowadays to model traffic on highways and to plan the battle operations of infantry with tanks; in these cases the smallest length scale is of the order of 10^2–10^4 cm. In modeling stellar conglomerations with a continuum approach, which is possible and often worthwhile, this scale (cf. Figure 1.10) is of the order of light years, 10^{17}–10^{18} cm.

1.2 Properties of a continuum and its motion. Density, flux and velocity. Law of mass balance

In order to model the equilibrium and motion of real bodies, their deformation, flow and fracture, we introduce an idealized medium, continuously filling a region in three-dimensional space. Now the goal is to determine laws governing the motion and, as a special case, the equilibrium of this idealized continuous medium. By this we mean:

(1) giving precise definitions of the quantitative properties of the motion in which we are interested; and
(2) obtaining mathematical expressions for these quantities as functions of position in space, time and other appropriate variables.

Let us first introduce several basic concepts, bearing in mind those readers who like formal definitions. A *body* is defined as a *three-dimensional manifold in a continuous medium*. We call the elements of this manifold *particles*. The region of three-dimensional space occupied by the body at a certain moment is called the *configuration of the body*. At different instants of time the body, generally speaking, has different configurations.

We emphasize the difference between a particle and a material point. *A particle in a continuous medium, in distinction to a material point, has all the properties of this medium.* Therefore its size, by a basic assumption, should be much larger than the characteristic length scale of the microstructure of a real body, although within the framework of continuum mechanics it is considered as infinitesimally small. Sometimes the term "physically infinitesimally small" is used. For instance, when modeling stellar conglomerations a particle can have a size of the order of light years (see e.g. Figures 1.7a, b), many orders of magnitude larger than the size of the bodies modeling terrestrial phenomena.

When introducing the concept of a continuous medium we meet the fundamental concept of *intermediate asymptotics*, which will accompany us throughout this entire book. As a reminder, by an asymptotic representation (an *asymptotics* for short) we mean the approximate representation of a function in a certain range of independent variables.

An intermediate asymptotics is determined, as already mentioned in the Introduction, in the following way. Assume that in a phenomenon under consideration there are two quantities, X_1 and X_2, each having the dimension of an independent variable x but of very different magnitudes:

$$X_1 \lll X_2 . \tag{1.2}$$

The notation $a \lll b$ means that there exists an interval of the values of the variable x such that $x \gg a$ but at the same time $x \ll b$. Then the asymptotic representation of certain properties of a phenomenon in the range of x

$$X_1 \ll x \ll X_2 ,$$

that is, for values large in comparison with X_1 but small in comparison with X_2, is called an intermediate asymptotics. More precisely, if in a problem we have two parameters X_1 and X_2 having the dimension of an independent variable x, we say that the intermediate asymptotics is an asymptotic representation for $x/X_1 \rightarrow \infty$ and $x/X_2 \rightarrow 0$. The concept of a particle in a continuous medium as being large in comparison with the medium's microscale but small in comparison with size of the body is a typical intermediate asymptotic concept.

To every body there corresponds some measure[2] called its mass, denoted by M. The limit

$$\rho = \lim \frac{M_n}{V_n} \tag{1.3}$$

for a sequence of bodies shrinking to a certain point (a particle) that have configurations Ω_n, masses M_n and volumes V_n is called the *density* of the continuous medium at the given point. In fact, we have in mind an "intermediate" limit in the sense mentioned above.

We come now to consider the properties of the motion of continua. We need to introduce the vital concept of an *observer*. We call an observer *a unit consisting of a reference system* (to start with, an inertial Newtonian system), *plus a clock and a system of units sufficient for measuring the properties of the class of phenomena under consideration* (e.g. the motion and equilibrium of bodies).

The motion of continua can be investigated in two ways. Either we can follow the motion of fixed particles of a body whose configuration was prescribed at a certain moment or we can follow the properties of the motion at points fixed in the reference system of an observer.

The first approach is known as the Lagrangian approach, and consists of the following. Let $\mathbf{X}$ be the radius vector of a particle of a body in the reference system of an observer at an initial moment $t = t_0$ and let $\mathbf{x}$ be the radius vector of the

---

[2] This measure can be introduced in various ways (for instance, in battle operations, the number of soldiers or tanks) but here we have in mind the ordinary Newtonian mass.

particle at an arbitrary time moment t. We are interested in the motion of fixed particles: using mathematical language, this motion is given by the mapping

$$\mathbf{x} = f(\mathbf{X}, t) . \tag{1.4}$$

The mapping $\mathbf{X} \to \mathbf{x}$ in (1.4) can be assumed to be continuous and single-valued and such that the inverse mapping $\mathbf{x} \to \mathbf{X}$ is also single-valued and continuous. In mathematics such mappings are called homeomorphisms. In the Lagrangian approach all other properties of the motion of a continuum are determined also as certain functions of the initial position of a particle, $\mathbf{X}$, in the reference system of an observer and of time t. For instance, the density ρ of a continuum is

$$\rho = \rho(\mathbf{X}, t) . \tag{1.5}$$

The second approach is named after Euler.[3] In the Eulerian approach, the properties of motion of a continuum are determined as functions of the current position $\mathbf{x}$ of a particle in the reference system of an observer and time t. Here the density is written as

$$\rho = \rho(\mathbf{x}, t) . \tag{1.6}$$

The velocity in the Eulerian approach is defined in a less direct way than in the Lagrangian approach. The definition of velocity is straightforward in the Lagrangian approach:

$$\mathbf{u} = \partial_t \mathbf{x}(\mathbf{X}, t) . \tag{1.7}$$

Indeed, the vector giving the initial position of a particle, $\mathbf{X}$, specifies the particle and does not depend on time t. At the same time, the vector $\mathbf{x}$ determines the current position of the particle at time t.

The definition of velocity in the Eulerian approach is made as follows. Let us take a particle O with radius vector $\mathbf{x}$ in the reference system of an observer. Let us construct at the point O an arbitrarily oriented elementary (infinitesimal) area $\Delta\Sigma$ having outward normal $\mathbf{n}$ and area $\Delta\Sigma$. Over an elementary time Δt, a mass ΔQ of the continuous medium passes through this area in the direction $\mathbf{n}$. Let us denote by j_n the density of flux of the continuum through the area having outward normal $\mathbf{n}$. This is equal to the limit of the ratio $\Delta Q/(\Delta\Sigma\Delta t)$ when the area shrinks to the point O (again, an intermediate asymptotic quantity). Of course, throughout this shrinking the diameter of the area is always considered to be much larger than the length scale of the microstructure, but we need not worry about that. Indeed, the concept of a continuous medium has already been introduced, and we are not

---

[3] In fact, the senior predecessor of Joseph-Louis Lagrange (Lagrange was Italian; his original given Italian name was Giuseppe Luigi) was Leonard Euler, who clearly knew both approaches. Therefore naming one particular approach as Lagrangian is a certain tribute to tradition. It is also a fact that Euler did so much that if we were to name everything he discovered after him it would mean his name would no longer be a distinction.

considering the motion of real matter but of a model of it, an idealized continuous medium. Now denote by j_1, j_2, j_3 the values of j_n for areas whose normals are directed along the coordinate axes of a Cartesian system x_1, x_2, x_3. We must determine what conditions are imposed on the properties of the motion just introduced by the law of mass conservation, i.e. by the fundamental condition that matter does not disappear or emerge from nowhere. Take an arbitrary finite region of three-dimensional space, Ω, bounded by a closed smooth surface $\partial\Omega$. We defined earlier the density ρ; according to this definition the mass in an elementary volume $d\omega$ is $\rho d\omega$, and the mass of the whole body having Ω as its configuration at time t is thus

$$\int_\Omega \rho(\mathbf{x}, t)\, d\omega . \tag{1.8}$$

The variation in this mass during the time interval between t and $t + dt$ is equal to

$$\left(\partial_t \int_\Omega \rho(\mathbf{x}, t)\, d\omega \right) dt . \tag{1.9}$$

This variation is composed, first of all, of an inflow of the continuous medium through the boundary $\partial\Omega$ into the region Ω during the elementary time interval. According to the definition given above of the flux density this inflow is

$$- dt \int_{\partial\Omega} j_n\, d\Sigma , \tag{1.10}$$

where $d\Sigma$ is an element of surface area and $\mathbf{n}$ is the unit vector of the outward normal to the surface $\partial\Omega$, hence the minus sign in formula (1.10).

Furthermore, the external inflow of matter into the volume Ω contributes

$$dt \int_\Omega Q\, d\omega \tag{1.11}$$

to the mass variation (1.9). Here $Q = Q(\mathbf{x}, t)$ is the density of mass inflow into the volume. The possibility of such an inflow for instance due to a chemical reaction or evaporation, should be, generally speaking, foreseen. By the law of mass conservation we have, equating (1.9) to the sum of (1.10) and (1.11),

$$\partial_t \int_\Omega \rho\, d\omega = \int_\Omega Q\, d\omega - \int_{\partial\Omega} j_n\, d\Sigma . \tag{1.12}$$

We remind the reader now of some elementary facts from vector and tensor calculus. Take two Cartesian orthonormal systems x_1, x_2, x_3 and x_1', x_2', x_3' with unit vectors $\mathbf{e}_1$, $\mathbf{e}_2$, $\mathbf{e}_3$ and $\mathbf{e}_1'$, $\mathbf{e}_2'$, $\mathbf{e}_3'$. Then the *geometric object* defined by the relations $\mathbf{a} = a_\alpha \mathbf{e}_\alpha$, $\mathbf{a} = a_\beta' \mathbf{e}_\beta'$ in each system and by the transformation law $a_i' = a_\gamma \lambda_{i\gamma}$, where $\lambda_{ij} = (\mathbf{e}_i' \cdot \mathbf{e}_j)$, is a first-rank tensor, i.e. a vector. Similarly, a geometric object $\boldsymbol{\Gamma}$ defined in each system by nine numbers Γ_{ij}, $\Gamma_{k\ell}'$ and the transformation law $\Gamma_{k\ell}' = \lambda_{k\beta} \lambda_{\ell\gamma} \Gamma_{\beta\gamma}$ is a second-rank tensor.

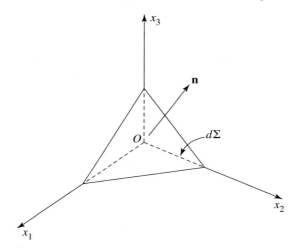

Figure 1.11 Neglecting terms of third order, we find that the flux over the surface of an elementary tetrahedron vanishes.

For two vectors $\mathbf{a}$ and $\mathbf{b}$, where $\mathbf{a} = a_\alpha \mathbf{e}_\alpha$, $\mathbf{b} = b_\beta \mathbf{e}_\beta$, three types of product are defined: the scalar, or inner, product $(\mathbf{a} \cdot \mathbf{b}) = a_\alpha b_\alpha$; the vector product defined as a determinant

$$[\mathbf{a}, \mathbf{b}] = \begin{vmatrix} \mathbf{e}_1 & \mathbf{e}_2 & \mathbf{e}_3 \\ a_1 & a_2 & a_3 \\ b_1 & b_2 & b_3 \end{vmatrix} ;$$

and the tensor product $\mathbf{a} \otimes \mathbf{b}$, i.e. the second-rank tensor $\mathbf{\Gamma}$ with components $\Gamma_{ij} = a_i b_j$ and transformation law $\Gamma'_{ij} = \Gamma_{\alpha\beta} \lambda_{i\beta} \lambda_{j\gamma}$. Note that only the scalar product is commutative: $(\mathbf{a} \cdot \mathbf{b}) = (\mathbf{b} \cdot \mathbf{a})$. The inner product $\mathbf{\Gamma} \cdot \mathbf{a}$ of the second-rank tensor $\mathbf{\Gamma}$ and the vector $\mathbf{a}$ is a vector with ith component $\Gamma_{i\alpha} a_\alpha$.

Now we can show that in fact j_n, the flux density through the elementary area having outward normal $\mathbf{n}$, is the projection onto $\mathbf{n}$ of a vector $\mathbf{j}$, which is naturally called the flux density vector of the continuum. Indeed, let us take Ω as an infinitesimally small tetrahedron (Figure 1.11) with three faces having normals opposite to the axes x_1, x_2, x_3 and the fourth face having area $d\Sigma$ and normal $\mathbf{n}$; the areas of the four faces constitute the elementary area of the boundary of the body, $\partial\Omega$. Integrals over this volume Ω will obviously be of the third order in smallness, whereas those over the surface will be of second order. Neglecting the former in comparison with the latter we obtain from (1.12)

$$\int_{\partial\Omega} j_n \, d\Sigma \simeq (j_1 n_1 + j_2 n_2 + j_3 n_3 - j_n) \, d\Sigma = 0 \,.$$

Here the n_i ($i = 1, 2, 3$) are the projections of the unit normal vector $\mathbf{n}$ onto the co-ordinate axes, so that the areas of the faces are $n_i\, d\Sigma$. Since the area $d\Sigma$ is arbitrary, we obtain

$$j_{\mathrm{n}} = j_\alpha n_\alpha , \qquad (1.13)$$

for an arbitrary direction of the normal $\mathbf{n}$. (Here and below we use the summation rule over repeated Greek indices (from 1 to 3), so that $j_\alpha n_\alpha = j_1 n_1 + j_2 n_2 + j_3 n_3$.)

Indeed, $j_\alpha n_\alpha = j'_\beta n'_\beta$ because the system of coordinates can be selected arbitrarily. Furthermore $\mathbf{n}$ is a vector, so that $n_\alpha = n'_\beta \lambda_{\beta\alpha}$. Therefore

$$j_\alpha n_\alpha - j'_\beta n'_\beta = n'_\beta (j_\alpha \lambda_{\beta\alpha} - j'_\beta) = 0 ,$$

and from this relation the vector transformation law for the components j_i follows immediately. The relation (1.13) means in particular that the quantity j_{n} defined above is the component parallel to the area vector $\mathbf{n}\, d\Sigma$ of the vector $\mathbf{j}(\mathbf{x}, t)$ having components j_1, j_2, j_3 in the orthogonal Cartesian system x_1, x_2, x_3: the vector law of transformation of the components follows from (1.13).

The *velocity vector* $\mathbf{u}(\mathbf{x}, t)$ of a particle $\mathbf{x}$ at time t is defined in the Eulerian approach as the ratio of the flux density of the continuous medium and its density:

$$\mathbf{u} = \frac{\mathbf{j}}{\rho} . \qquad (1.14)$$

We assume that the density, flux density and velocity are smooth functions, i.e. that *they all have the necessary time and space coordinate derivatives*.[4]

Transforming the last integral on the right-hand side of (1.12) into an integral over volume (using Gauss' theorem) and taking into account (1.14), we can reduce (1.12) to the form

$$\int_\Omega (\partial_t \rho + \nabla \cdot \rho\mathbf{u} - Q)\, d\omega = 0 . \qquad (1.15)$$

Here ∇ is the Hamiltonian vectorial operator "nabla", the vector $\nabla = \mathbf{e}_\alpha \partial_\alpha$ where $\mathbf{e}_\alpha$ are the unit vectors of the coordinate axes. For a scalar f, the vector $\nabla \cdot f = \mathbf{e}_\alpha \partial_\alpha f = \operatorname{grad} f$ is called the *gradient* of f. For a vector $\mathbf{q}$, the scalar $\nabla \cdot \mathbf{q} = \partial_\alpha q_\alpha = \operatorname{div} \mathbf{q}$ is called the *divergence* of $\mathbf{q}$. Furthermore the vector $[\nabla, \mathbf{q}] = \operatorname{curl} \mathbf{q}$ is called the *curl* or the *vorticity* of $\mathbf{q}$.

The space region Ω is arbitrary; therefore we obtain

$$\partial_t \rho + \nabla \cdot \mathbf{j} = Q \quad \text{or} \quad \partial_t \rho + \nabla \cdot (\rho\mathbf{u}) = Q . \qquad (1.16)$$

---

[4] There exists a large class of motions, for instance the motion of gas with large velocities, where such an assumption, generally speaking, does not hold everywhere in the domain where the continuum motion is considered: the functions ρ, $\mathbf{j}$, $\mathbf{u}$ could have discontinuities. We will see in Chapter 10 how to deal with such cases.

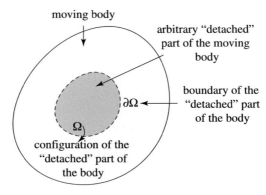

Figure 1.12 There is a momentum flow through the boundary of the configuration of the "detached" part of the body. Also, the momentum of this part is influenced by the ambient part of the body.

In the frequently considered case when there is no inflow or outflow of matter we have $Q = 0$ and the equation of mass conservation takes the form

$$\partial_t \rho + \nabla \cdot \mathbf{j} = 0 \quad \text{or} \quad \partial_t \rho + \nabla \cdot (\rho \mathbf{u}) = 0 , \qquad (1.17)$$

which is known as the *continuity equation*. Introduce in the observer's reference system the fixed orthonormal Cartesian coordinates x_i ($i = 1, 2, 3$). In these coordinates, equation (1.17) takes the form

$$\partial_t \rho + \partial_\alpha \rho u_\alpha = 0 . \qquad (1.18)$$

Here u_i ($i = 1, 2, 3$) are the components of the velocity vector $\mathbf{u}$ in the coordinate system x_i. We remind the reader of the summation rule over repeated Greek indices.

We note that the equation of mass conservation (1.18), which we have obtained with the Euler approach, is not closed. It relates four quantities: the density ρ and the three components u_i of the velocity vector. Therefore we will try to overcome this difficulty by considering another conservation law, the law of momentum balance.

1.3 Law of momentum balance. Stress tensor

Let us use the Euler approach. Consider an arbitrary region Ω, fixed in time and bounded by a smooth surface $\partial\Omega$, in three-dimensional space occupied by a moving continuous medium (Figure 1.12). The momentum of a body enclosed in any elementary volume $d\omega$ is equal to $\rho \mathbf{u} \, d\omega$, so that the momentum of an arbitrarily detached part of a body at time t having the region Ω as its configuration is equal to

$$\int_\Omega \rho \mathbf{u} \, d\omega \,. \tag{1.19}$$

Its variation in the time interval from t to $t + dt$ is, to first order,

$$\left(\partial_t \int_\Omega \rho \mathbf{u} \, d\omega \right) dt \,. \tag{1.20}$$

Evidently this variation has two parts: first, the momentum inflow from mass force which acts on the body and, second, the momentum inflow through the boundary of the body, $\partial \Omega$. By definition, the mass force acting on an elementary volume $d\omega$ is equal to $\rho \mathbf{F} \, d\omega$, where $\mathbf{F}$ is a vector that depends, generally speaking, on the position of the element and on the time t. (The simplest examples of mass forces are gravity and centrifugal forces.) Thus, the inflow of momentum to the body under consideration due to the mass force in the time dt is equal to

$$dt \int_\Omega \rho \mathbf{F} \, d\omega \,. \tag{1.21}$$

Consider now the momentum influx through the boundary $\partial \Omega$. Assume for the time being that the boundary of the detached part of the body $\partial \Omega$ is free: the ambient continuous medium does not act on it. A mass $(\mathbf{j} \cdot \mathbf{n}) \, d\Sigma = (\rho \mathbf{u} \cdot \mathbf{n}) \, d\Sigma$ flows per unit time through elementary area $d\Sigma$ of the surface $\partial \Omega$, carrying with it amount of momentum equal to $(\rho \mathbf{u} \cdot \mathbf{n})\mathbf{u} \, d\Sigma$. Therefore the variation of the body's momentum over time dt due to the inflow of momentum through the boundary $\partial \Omega$ is equal to

$$-dt \left(\int_{\partial \Omega} (\rho \mathbf{u} \cdot \mathbf{n})\mathbf{u} \, d\Sigma \right) ;$$

again, a minus sign appears because $\mathbf{n}$ is an outward normal.

Note now that the quantity $(\rho \mathbf{u} \cdot \mathbf{n})\mathbf{u}$ is a vector with ith component $\rho u_\alpha n_\alpha u_i$. It is equal to the product of $\rho \mathbf{u} \otimes \mathbf{u}$, the tensor product of the two vectors $\rho \mathbf{u}$ and $\mathbf{u}$, having components $\rho u_\alpha u_i$ and the vector $\mathbf{n}$. Therefore the previous expression can be written in the form

$$- dt \left(\int_{\partial \Omega} (\rho \mathbf{u} \otimes \mathbf{u}) \cdot \mathbf{n} \, d\omega \right). \tag{1.22}$$

However, the boundary $\partial \Omega$ of the body is arbitrary and, therefore, in general not free. There is a force on the body due to the action of the remaining part of the continuous medium. Therefore the variation in the momentum of the body having configuration Ω due to the inflow of momentum through the boundary $\partial \Omega$ can be written in the form

$$- dt \left(\int_{\partial \Omega} (\rho \mathbf{u} \otimes \mathbf{u} - \boldsymbol{\sigma}) \cdot \mathbf{n} \, d\Sigma \right). \tag{1.23}$$

Here σ is an additional term expressing the force action of the ambient continuous medium on the surface $\partial\Omega$ of the body with configuration Ω.

As a reminder, the tensor product of vectors $\rho\mathbf{u}$ and $\mathbf{u}$ is a tensor of the second rank with components $\rho u_i u_j$. Therefore, owing to the invariance principle, the *quantity σ must also be a second-rank tensor*, and we denote its components by σ_{ij}. Otherwise the expression (1.23) would be non-invariant with respect to a transformation of coordinates (for instance, a transition from one orthonormal Cartesian system to another), and this would violate the principle of the equivalence of observers, which is impossible for a quantity having a physical sense. The second-rank tensor σ which determines the internal forces acting in a continuous medium is called the *stress tensor*. From (1.20), (1.21) and (1.23) we obtain

$$\partial_t \int_\Omega \rho\mathbf{u}\,d\omega = \int_\Omega \rho\mathbf{F}\,d\omega - \int_{\partial\Omega} (\rho\mathbf{u}\otimes\mathbf{u} - \sigma)\cdot\mathbf{n}\,d\Sigma\,. \tag{1.24}$$

Using the same assumption as in the derivation of the equation of mass balance, namely, that density, velocity and the stress tensor are smooth functions, we can transform the surface integral on the right-hand side of (1.24) into a volume integral:

$$\int_\Omega [\partial_t(\rho\mathbf{u}) + \nabla\cdot(\rho\mathbf{u}\otimes\mathbf{u} - \sigma) - \rho\mathbf{F}]\,d\omega = 0\,. \tag{1.25}$$

As before, ∇ is the vectorial operator with components $\mathbf{e}_i\partial_i$. Bearing in mind that the body configuration Ω is arbitrary, we obtain from (1.25) the differential equation for momentum balance,

$$\partial_t\rho\mathbf{u} + \nabla\cdot(\rho\mathbf{u}\otimes\mathbf{u} - \sigma) = \rho\mathbf{F}\,. \tag{1.26}$$

In the orthonormal Cartesian coordinate system x_1, x_2, x_3, equation (1.26) takes the form

$$\partial_t\rho u_i + \partial_\alpha(\rho u_\alpha u_i - \sigma_{i\alpha}) = \rho F_i\,. \tag{1.27}$$

The tensor $\rho\mathbf{u}\otimes\mathbf{u}$ has components $\rho u_i u_j$, so that its divergence $\nabla\cdot\rho\mathbf{u}\otimes\mathbf{u}$ is a vector with ith component $\partial_\alpha\rho u_i u_\alpha = u_i\partial_\alpha\rho u_\alpha + \rho u_\alpha\partial_\alpha u_i$. The last term is just the ith component of the vector $(\rho\mathbf{u}\cdot\nabla)\cdot\mathbf{u}$. Using the continuity equation (1.17) it is possible to represent the momentum balance equation (1.26) in the form

$$\partial_t\mathbf{u} + (\mathbf{u}\cdot\nabla)\mathbf{u} = \mathbf{F} + \frac{1}{\rho}\nabla\cdot\sigma\,. \tag{1.28}$$

The expression on the left-hand side of equation (1.28) is the "individual" time derivative of the velocity, i.e. the time derivative calculated for a fixed particle whose position changes with time, not the time derivative for a given point in space.

The individual (or "total") derivative is denoted by d/dt and defined by the relation

$$\frac{d}{dt} = \partial_t + (\mathbf{u} \cdot \nabla), \tag{1.29}$$

so that equation (1.28) can be rewritten in the form

$$\frac{d\mathbf{u}}{dt} = \mathbf{F} + \frac{1}{\rho} \nabla \cdot \sigma. \tag{1.30}$$

Let us clarify the physical meaning of the stress tensor σ. Taking into account the continuity equation, equation (1.24) can be rewritten in the form

$$\int_\Omega \rho \frac{d\mathbf{u}}{dt} d\omega = \int_\Omega \rho \mathbf{F} d\omega + \int_{\partial\Omega} (\sigma \cdot \mathbf{n}) d\Sigma. \tag{1.31}$$

The left-hand side of this relation is the rate of variation of the momentum of the body having configuration Ω at time t. The first term on the right-hand side of (1.31) is the total mass force acting on the body. Evidently the second term on the right-hand side of (1.31), which is a surface integral, represents the bulk force acting on the body from the side of the ambient continuous medium through its boundary, the surface $\partial\Omega$. This force is the sum of the elementary forces acting on an element of the surface having outward normal $\mathbf{n}$ and area $d\Sigma$; the elementary force acting on such an elementary surface is proportional to $d\Sigma$ and equal to

$$(\sigma \cdot \mathbf{n}) d\Sigma = \mathbf{p} d\Sigma. \tag{1.32}$$

Thus the ith component p_i of the force vector $\mathbf{p}$ acting per unit area on the elementary area $d\Sigma$ with external normal $\mathbf{n}$ is given by

$$p_i = \sigma_{i\alpha} n_\alpha, \tag{1.33}$$

because, by definition of a second-rank tensor, the quantity $\sigma \cdot \mathbf{n}$ is just the vector with components $\sigma_{i\alpha} n_\alpha$.

Evidently, the component σ_{ij} of the stress tensor σ is, as (1.33) shows, the ith component of the force per unit area on the area element $d\Sigma$ having normal $\mathbf{n}$ directed along the axis x_j.

We now consider several examples. Let a long cylindrical rod with cross-sectional area S be stretched in the direction of its axis by a force P. Take the direction of the axis of the rod as the x_1 axis, with axes x_2 and x_3 orthogonal to it within a certain cross-section of the rod (it is immaterial which cross-section if it is far from the supports gripping the rod). We easily obtain

$$\sigma_{11} = \frac{P}{S}, \qquad \sigma_{22} = \sigma_{33} = \sigma_{12} = \sigma_{13} = \sigma_{31} = \sigma_{21} = \sigma_{23} = \sigma_{32} = 0. \tag{1.34}$$

Another example: at an arbitrary area element in a fluid at rest, the force (in this case the pressure) is always directed along the normal to the area and its magnitude does not depend on the direction of the normal (this is Pascal's law). Thus

$$\mathbf{p} = -p\mathbf{n} \tag{1.35}$$

and we obtain from (1.35) and (1.32) that

$$\boldsymbol{\sigma} = -p\mathbf{I} \, . \tag{1.36}$$

Here p is a scalar, the pressure, and $\mathbf{I}$ is a unit tensor of second rank having components δ_{ij}, where $\delta_{ij} = 1$ for $i = j$ and $\delta_{ij} = 0$ otherwise.

A final example: the shear stress in Newton's experiment establishing the basic law of viscous flow (Figure 2b in the introductory chapter), which is given by $\sigma_{12} = \eta V/h$.

We come to a conclusion which seems to be disappointing: the additional balance equation, the equation of momentum balance, does not reduce the degree of non-closedness of the system. Indeed, the four equations (1.18) and (1.27) have 13 unknowns, the density ρ, the three components of velocity u_1, u_2, u_3 and the nine components of the stress tensor σ_{ij}. To continue in the same way, employing new balance equations, is hopeless, and so we will select a different path. But, before that, in the next chapter, we will present a very important tool that will be used widely throughout the book.

2
Dimensional analysis and physical similitude

In the first chapter we introduced the important concept of an observer and formulated the invariance principle, which states the equivalence (equal rights in mathematical modeling!) of observers. In this chapter we consider what follows from the equivalence of observers whose units of measurement are of the same physical nature but of different magnitudes.

This looks simple but in this case the consequences of the invariance principle are far from trivial. Indeed, we will show that it follows from the principle of equivalence of observers that the functions describing physical laws have a fundamental property which is called *generalized homogeneity*. This property allows a reduction in the numbers of arguments of these functions and simplifies their determination in a numerical computation or in an experiment. The corresponding procedure is called *dimensional analysis*. Dimensional analysis is closely related to the rules of the modeling of physical phenomena, which make up the essence of the *theory of physical similitude*. Dimensional analysis and the theory of similitude will be presented in this chapter in sufficient detail for their use throughout the whole book. More detailed presentation of the subject of this chapter can be found in the author's book (Barenblatt, 2003).

2.1 Examples

Example 1. In the autumn of 1940, when the development of atomic weapons was beginning, a fundamental question arose concerning the mechanical action of the energy released during an atomic explosion. An outstanding American expert in explosives, G. B. Kistyakovsky, reported that even if such a weapon were created all its energy would go to radiation and would have essentially no mechanical effect.

The problem of the mechanical effect of concentrated energy release was considered by eminent researchers, G. I. Taylor in Great Britain and J. von Neumann

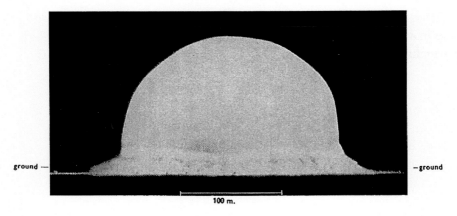

ground —   —ground

100 m.

Figure 2.1 Photograph of the fireball of the atomic explosion in New Mexico at $t = 15$ ms, confirming in general the spherical symmetry of the motion of the gas. From (Taylor, 1950) by permission of the Royal Society.

in the United States. The solution was obtained by them virtually simultaneously (Taylor (1941) presented his report on Friday, 27 June 1941; von Neumann (1941) took it home for the weekend, apparently to check the calculations again, and presented his work on Monday, 30 June 1941). Solving a corresponding problem by integrating the equations of gas dynamics (greatly simplified by using arguments which will be presented below), they obtained that the radius of the shock-wave front increases according to the law

$$r_{\mathrm{f}} = 0.89 \left(\frac{Et^2}{\rho_0} \right)^{1/5}. \tag{2.1}$$

Both G. I. Taylor and J. von Neumann assumed the explosion to be spherically symmetric. Subsequent testing at Alamogordo, New Mexico, on 16 July 1945 confirmed this assumption (see Figure 2.1). We will now obtain the relation (2.1) up to the numerical factor, using arguments based on the equivalence of observers having units of measurement of differing magnitudes.

First we ask the following question: what are the quantities upon which the radius of the shock wave depends? It is natural to assume that if the energy is released very rapidly or even instantaneously within a very small region, one might say at a point, then the radius of the front depends on the energy E but does not depend on the time interval during which the energy was released or on the size of the device. It is clear furthermore that the radius r_{f} of the front depends on the time t after the explosion and on the initial density of the air ρ_0: there are no shock waves in a vacuum. At the initial stage of the explosion the shock wave is very intense; therefore the pressure just behind the shock-wave front exceeds the air pressure before the front by many orders of magnitude, and it is natural to assume that the air pressure ahead of the front does not influence the shock front propagation. Therefore

$$r_f = f(E, t, \rho_0) \,. \tag{2.2}$$

Let the first observer possess units of measurement of length, mass and time; to be definite let us say 1 cm, 1 g and 1 s. All other observers, whose equivalence with the first one is stated by the invariance principle, have corresponding units of measurement of length, mass and time which can be represented in the form

$$\frac{\text{cm}}{L} \,, \qquad \frac{\text{g}}{M} \,, \qquad \frac{\text{s}}{T} \,. \tag{2.3}$$

Here L, M, T are *arbitrary positive numbers*. Therefore the numbers expressing time for these general observers are a factor T larger than those for the first observer; the numbers expressing mass for the general observers are a factor M larger and the numbers expressing length are a factor L larger.

The units of length, time and mass are *fundamental* units. The unit of density is a *derived* unit. By definition it is the density of a homogeneous body containing a unit of mass in a unit of volume. Therefore passing from the first observer to the general observer we decrease the unit of density ML^{-3} times, and all densities for general observers have numerical values ML^{-3} times larger than for the first. In the same way, when passing from the first to the general observer, numerical values of all physical quantities vary. The numerical factor that expresses this variation is determined by the dimension function, or simply *dimension* of the physical quantity under consideration.[1] For instance, in passing from the first to the general observer we increase values of all lengths by a factor L. We say the dimension of length is L. Similarly, the dimension of mass is M, the dimension of time is T, the dimension of density is ML^{-3}, the dimension of velocity is LT^{-1} and that of acceleration LT^{-2}. We emphasize again that L, M, T are positive numbers like 3, $\sqrt{2}$ or π, and nothing more.

Furthermore, force is related to mass and acceleration by Newton's law: it is equal to the product of mass and acceleration. The dimensions of both sides of every relationship having a physical content should be equal. Indeed, if they were different then equality for one observer would become an inequality for a different observer. Thus the dimension of force is equal to the dimension of the product of mass by acceleration, i.e. MLT^{-2}. The unit of energy is also a derived unit; it is equal to the product of a unit force by a unit length. Thus the dimension of energy is equal to ML^2T^{-2}.

---

[1] The dimension function is always a power monomial composed from arguments such as M, L, T (see the proof in e.g. Barenblatt (2003), pp. 18–20).

Now, let us construct from the quantities E, t, ρ_0, which are the arguments of the function f in (2.2), the combination Et^2/ρ_0. It is easy to verify that this combination has dimension L^5: it increases when passing from the first observer to the general observer by a factor L^5. Therefore the quantity

$$\Pi = r_{\mathrm{f}}(Et^2/\rho_0)^{-1/5} \tag{2.4}$$

remains invariant when passing from the first observer to the general observer. Quantities which, like Π, remain invariant when the magnitudes (but not the physical nature) of the fundamental units vary are called *dimensionless*. Their dimension is equal to unity. All other quantities are called *dimensional*. To each dimensional quantity there corresponds a *dimension*, a function different from unity, which determines the variation of the numerical value of this quantity when one is passing from the first observer to the general observer. Following J. C. Maxwell, the dimension of a physical quantity a is denoted $[a]$.

Obviously the relation (2.2) can be rewritten in the form

$$\Pi = F\left(\left(\frac{Et^2}{\rho_0}\right)^{1/5}, t, \rho_0\right) . \tag{2.5}$$

We remind the reader that the arguments of the function F in (2.5) are numbers related to a certain system of units.

Let us pass now from the system of units of the first observer to another observer, having a unit of mass M times less (where M is an arbitrary positive number) and the same units of length and time. Then the number ρ_0 will increase M times, and numbers Π (dimensionless), $(Et^2\rho_0)^{1/5}$ (having the dimension of length), and t (having the dimension of time) remain the same. But this means that the function F in (2.5) remains invariant when its third argument, ρ_0, varies arbitrarily and the first two arguments remain invariant, in other words, that F is independent of ρ_0. Furthermore, let us pass to an observer whose units of length and mass remain the same, but whose unit of time decreases arbitrarily by a factor T. In this case the first argument of the function F remains invariant, as well as the function F itself, whereas the second argument varies arbitrarily by the factor T. This means that the function F does not depend on t either.

Finally, let us pass to an observer whose unit of length is decreased L times, the other units remaining the same. The value of the function F will remain invariant whereas its first argument varies arbitrarily. This means that the function F also does not depend on the first argument, i.e. that the function F is in fact a constant:

$$\Pi = \frac{r_{\mathrm{f}}}{(Et^2/\rho_0)^{1/5}} = \mathrm{const} . \tag{2.6}$$

From this one obtains the relation

$$r_{\mathrm{f}} = \mathrm{const} \times (Et^2/\rho_0)^{1/5} . \tag{2.7}$$

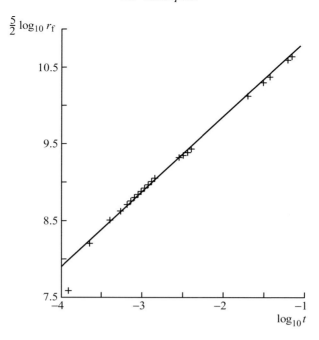

Figure 2.2 Logarithmic plot of the fireball radius, showing that $r_f^{5/2}$ is proportional to time t. From (Taylor, 1950) by permission of the Royal Society.

The constant is obtained by solving the equations of gas dynamics, also simplified by dimensional analysis (see e.g. Barenblatt (1996), pp. 80–4), and happens to be nearly equal to unity: it is ≈ 0.89. Formula (2.1) shows that if the radius of the shock-wave front r_f is measured at different times, experimental points in the logarithmic coordinates $\log t$, $(5/2) \log r_f$ should collapse onto the straight line

$$\frac{5}{2} \log r_f = \frac{1}{2} \log \frac{E}{\rho_0} + \log t + \frac{5}{2} \log \text{const} \tag{2.8}$$

having a slope equal to unity. This was triumphantly confirmed by G. I. Taylor (Taylor, 1950) who processed data from a series of high-speed photographs (by J. Mack) of the fireball expansion, taken four years before, during the first American nuclear test (Figure 2.1). Moreover, it was possible to determine the energy of the explosion from the experimental data (the photographs) presenting the dependence of the radius of the shock wave on the time elapsed; it could be deduced from the ordinate intercept of the straight line (2.8) constructed from the experimental points (Figure 2.2). At the time, Taylor's publication of this value, which turned out to be approximately equivalent to 20 000 tons of conventional TNT explosive, caused, in his words, "much embarrassment" in American government circles: this figure was considered top secret, whereas Mack's film was not classified.

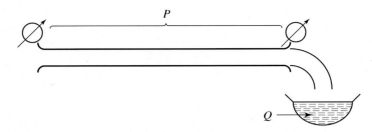

Figure 2.3 A schematic diagram showing the experiments of Bose, Rauert and Bose. The time τ required to fill a vessel of volume Q and the pressure drop between the ends of the pipe P were measured for the steady turbulent flow of various fluids through the pipe. From Barenblatt (2003).

Example 2: At the beginning of the twentieth century, the physical chemists E. Bose, D. Rauert and M. Bose published a series of experimental studies of internal turbulent friction in various fluids. The experiments were carried out in the following way (Figure 2.3). Various fluids (water, chloroform, bromoform, mercury, ethyl alcohol etc.) were allowed to flow through a pipe in a regime of steady (of course, on average) turbulence. The time τ required to fill a vessel with a fixed volume Q and the pressure drop P between the ends of the pipe were measured. As was customary, the results of the measurements were represented in the form of a series of tables and curves (similar to those in Figure 2.4), showing the pressure drop P as a function of the filling time τ.

At this time, the work of Bose, Rauert and Bose attracted the attention of Th. von Kármán (von Kármán, 1957), then a young researcher, later to become one of the greatest figures in applied mechanics of the twentieth century. He also considered quantities used by experimentalists that were not very common and subjected their results to a specific analysis, which can be represented in the following way. First, he noticed that the pressure drop between the ends of the pipe P depends on the time τ required for the vessel to be filled and its on volume Q as well as on properties of the fluid, its viscosity coefficient η and its density ρ:

$$P = f(\tau, Q, \eta, \rho) . \tag{2.9}$$

Let us demonstrate the dimensions of the quantities which enter relation (2.9). By definition, the pressure is a force divided by an area. Therefore the dimension of pressure is equal to $MLT^{-2}/L^2 = ML^{-1}T^{-2}$. The dimension of the filling time τ is T, the dimension of volume Q is L^3 and the dimension of density, as was shown before, is equal to ML^{-3}. To obtain the dimension of the coefficient of viscosity η we have to remember the definition of this coefficient. According to Newton's formula (see the introduction), known from high-school physics, if a fluid layer of thickness h is contained between two plates, one of which is moving parallel to the

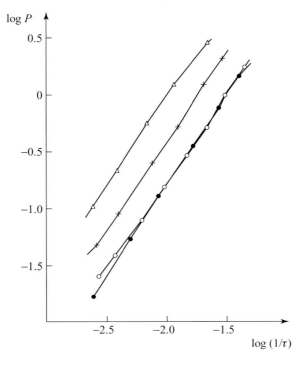

Figure 2.4 The experimental results of Bose, Rauert and Bose in their original form: ∘, water; •, chloroform; +, bromoform; Δ, mercury (P is in kgf/cm^2 and τ in seconds). The curves are different for different fluids. From von Kármán (1957).

other with velocity V whereas the other is at rest, a shear force σ acts on unit area of the plates. This force is equal to

$$\sigma = \eta V/h , \qquad (2.10)$$

where η is the coefficient of viscosity, a property of the fluid.

The dimension of σ is equal to the dimension of pressure. The dimensions of both sides of (2.10) should be equal; therefore the dimension of the coefficient of viscosity η is equal to $ML^{-1}T^{-1}$.

Note that the dimensions of the first three arguments of the function f in (2.9) are *independent*. This means that by varying the units of measurement of time, length, and mass we can vary the numerical values of one of these arguments, leaving the other two invariant. Indeed, by varying the unit of time we can vary arbitrarily the first argument in (2.9), leaving invariant the values of the second and third arguments (for the third argument, by a compensating variation in the unit of mass). This reasoning, however, is correct neither for the pressure P nor for the density ρ. Indeed, as is easy to verify,

$$[P] = \frac{M}{LT^2} = [\tau]^{-1}[Q]^0[\eta] \,,$$

$$[\rho] = \frac{M}{L^3} = [\tau][Q]^{-2/3}[\eta] \,.$$

(2.11)

Therefore, when we are arbitrarily varying the units of measurement of the basic quantities time, length and mass, we cannot leave the values of quantities P and ρ invariant. Let us form the *dimensionless* quantities

$$\Pi = \frac{P}{\eta\tau^{-1}} \,, \qquad \Pi_1 = \frac{\rho}{\tau Q^{-2/3}\eta} \,.$$

(2.12)

It is permissible to rewrite (2.9) in the form

$$\Pi = F(\tau, Q, \eta, \Pi_1) \,.$$

(2.13)

Now reduce arbitrarily the unit of mass by a factor M, leaving the units of length and time invariant: then the numerical value of η will increase M times whereas the numerical values of the dimensional quantities τ and Q, as well as the numerical values of the dimensionless quantities Π and Π_1, will remain invariant. But, as in the previous example, this means that the function F does not depend upon η. Furthermore, by arbitrarily varying the unit of time we obtain that only the argument τ varies, leaving Q, Π and Π_1 invariant. This means that function F does not depend upon τ. Finally, arbitrarily varying the unit of length, we can obtain in the same way that the function F does not depend upon Q, so that we can write

$$\Pi = \Phi(\Pi_1) \,.$$

(2.14)

However, further simplification is impossible: the quantities Π, Π_1 are dimensionless and when we pass from the first observer to the general observer their numerical values remain invariant. Even so, the result obtained is impressive: returning to the original dimensional variables by means of (2.12), equation (2.14) becomes

$$P = \frac{\eta}{\tau} \Phi\left(\frac{\rho}{\eta\tau Q^{-2/3}}\right).$$

(2.15)

Thus instead of determining the function of four variables $f(\tau, Q, \eta, \rho)$ according to relation (2.9), a function Φ of only one variable needs to be determined. Formula (2.15) shows that if the experimental data are plotted in the coordinates $\rho/(\eta\tau Q^{-2/3})$ and $P/(\eta\tau^{-1})$ then the experimental points should collapse to a single curve; von Kármán processed the experimental data of Bose, Rauert and Bose and confirmed this fact decisively (Figure 2.5). It is clear that a similar analysis

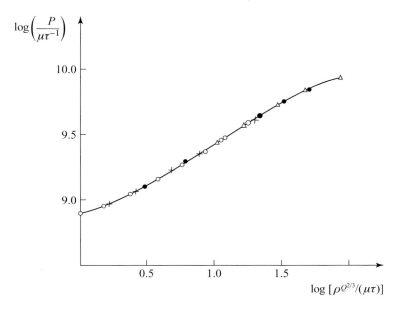

Figure 2.5 The experimental results of Bose, Rauert and Bose as represented by von Kármán, who used dimensional analysis. All experimental points lie on a single curve. From von Kármán (1957).

performed beforehand could have reduced significantly the amount of experimental work required.

2.2 Dimensional analysis

The examples presented in the previous section demonstrated the general recipe for the application of *dimensional analysis*. Superficially it looks very simple; the whole process can be reduced to a certain sequence of steps.

Suppose that we are interested in a property a of some phenomenon (there may be many such properties). We proceed as follows.

(1) We establish the *governing parameters* $a_1, \ldots, a_n$ upon which quantity a depends; thus

$$a = f(a_1, \ldots, a_n) . \tag{2.16}$$

All these quantities $a, a_1, \ldots, a_n$ are, generally speaking, dimensional, although some of them may be dimensionless.
(2) We choose from the system $a_1, \ldots, a_n$ those parameters whose dimensions are independent. This means their numerical values can vary independently when the basic units of measurement, such as time, length and mass, are changed. We

denote these parameters as $a_1, \ldots, a_k$. It is clear that the number k of parameters with independent dimensions can vary from zero, when $a_1, \ldots, a_n$ are all dimensionless, to the full number n of governing parameters, as in the first example in Section 2.1. The dimensions of the remaining parameters $a_{k+1}, \ldots, a_n$ and of the quantity a to be determined can be expressed as products of powers of the dimensions of the governing parameters with independent dimensions, $a_1, \ldots, a_k$:

$$
\begin{aligned}
[a] &= [a_1]^p \cdots [a_k]^r , \\
[a_{k+1}] &= [a_1]^{p_{k+1}} \cdots [a_k]^{r_{k+1}} , \\
&\vdots \\
[a_n] &= [a_1]^{p_n} \cdots [a_k]^{r_n} .
\end{aligned}
\tag{2.17}
$$

(3) The relationship (2.16) under consideration is now represented in the form of a relationship between dimensionless quantities:

$$
\Pi = \Phi(\Pi_1, \ldots, \Pi_{n-k}) .
\tag{2.18}
$$

Here the dimensionless parameters $\Pi, \Pi_1, \ldots, \Pi_{n-k}$ are introduced according to the relations

$$
\begin{aligned}
\Pi &= \frac{a}{a_1^p \cdots a_k^r} , \\
\Pi_1 &= \frac{a_{k+1}}{a_1^{p_{k+1}} \cdots a_k^{r_{k+1}}} , \\
&\vdots \\
\Pi_{n-k} &= \frac{a_n}{a_1^{p_n} \cdots a_k^{r_n}} .
\end{aligned}
\tag{2.19}
$$

The numbers $p, r, \ldots, p_{k+1}, \ldots, r_{k+1}, p_n, \ldots, r_n$ can be found through a simple calculation comparing the dimensions of the corresponding parameters. We demonstrated how to do this in the examples in Section 2.1. It is clear that the governing parameters with independent dimensions cannot be arguments of the function Φ, since all these arguments must be dimensionless.

Dimensional analysis assists us in the following way: the number of dimensionless parameters is reduced by the number of parameters with independent dimensions. Indeed, we obtain from relations (2.18), (2.19):

$$
a = \Pi a_1^p \cdots a_k^r = a_1^p \cdots a_k^r \, \Phi(\Pi_1, \ldots, \Pi_{n-k}) .
\tag{2.20}
$$

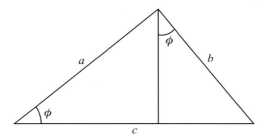

Figure 2.6 A proof of Pythagoras' theorem using dimensional analysis. From Barenblatt (2003).

This means that, in relation (2.16), the function $f(a_1, \ldots, a_n)$ which represents the physical law in which we are interested has the fundamental property of *generalized homogeneity*:

$$f(a_1, \ldots, a_n) = a_1^p \cdots a_k^r \, \Phi\left(\frac{a_{k+1}}{a_1^{p_{k+1}} \cdots a_k^{r_{k+1}}}, \ldots, \frac{a_n}{a_1^{p_n} \cdots a_k^{r_n}} \right). \qquad (2.21)$$

This expression, which is the basic statement of dimensional analysis, is sometimes called the Π-*theorem*. It simplifies investigations: instead of having to determine a function of n arguments, a researcher can determine a function of $n - k$ arguments. Roughly speaking, *the amount of work required to determine the desired function is reduced by as many orders of magnitude as there are governing parameters with independent dimensions.*

In conclusion, we consider a rather amusing example of the application of dimensional analysis. We shall "prove" Pythagoras' theorem.

The area of a right-angled triangle, S_c (Figure 2.6), is completely determined by its hypotenuse c and, for definiteness, the smaller of its acute angles ϕ. Thus

$$S_c = f(c, \phi).$$

The angle ϕ is dimensionless, and it is clear that in this example $n = 2$ and $k = 1$. Dimensional analysis gives, using (2.19),

$$\Pi = \frac{S_c}{c^2} = \Phi(\Pi_1), \quad \Pi_1 = \phi, \quad S_c = c^2 \Phi(\phi).$$

The altitude perpendicular to the hypotenuse of the triangle divides it into two similar right-angled triangles (Figure 2.6) with hypotenuses respectively equal to the sides a and b of the large triangle. The previous equation yields the following result for the areas of these triangles:

$$S_a = a^2 \Phi(\phi), \quad S_b = b^2 \Phi(\phi),$$

where $\Phi(\phi)$ is the same function as for the larger triangle.

The sum of the areas S_a and S_b of the two smaller triangles is equal to the area of the basic triangle S_c, so that

$$c^2\Phi(\phi) = a^2\Phi(\phi) + b^2\Phi(\phi) \, .$$

Cancelling out $\Phi(\phi)$ in the latter relation we find that

$$c^2 = a^2 + b^2,$$

which is the desired result.

The examples discussed above demonstrate that the seemingly trivial procedure of dimensional analysis can lead to very strong results. However, this seeming simplicity is in fact illusory. Indeed, this procedure is very effective where the process is reduced finally to determining a constant (see Example 1 in Section 2.1) or a function of a single dimensionless variable (see Example 2 in Section 2.1). Therefore correctly choosing the set of governing parameters becomes the most important factor. It is important not only to take all essential parameters into account but also to avoid including superfluous ones! The correct choice, especially if the exact mathematical formulation is not known, requires good intuition and a great deal of attention to the qualitative analysis of the phenomenon under consideration. Therefore the central point is not the trivial application of the formal procedure of dimensional analysis but, rather, the correct choice of parameters governing the quantities in which we are interested. We will return to this point repeatedly in this book.

2.3 Physical similitude

2.3.1 Modeling

Before a large expensive object (for example, a ship or aeroplane) is constructed, experiments on *models* are needed to determine the best properties of the object for future operating conditions. Many different kinds of measurement are carried out on models. For example, the lift and drag of an aeroplane model as air flows past it can be measured in a wind tunnel, as can aerodynamic loading that might cause a television tower to collapse, etc. Clearly one must know how to scale (i.e. re-evaluate) the results of experiments carried out on a model up to the full-scale object in question. If one does not know how to do this, modeling is a useless pursuit.[2] The concept of *physically similar phenomena* is central to correct modeling.

---

[2] A little parable for illustration: A man asks God how Mount Everest seems to him. "Like a grain of sand does to you." "And how about a billion pounds?" "Like a ha'penny." "So, God, please give me a billion pounds." "Wait a second please." (A Galactic second in God's time scale is one hundred thousand years.)

This concept of physical similitude is a natural generalization of the concept of similarity in geometry. For example, two triangles are similar if they differ only in the numerical values of their dimensional parameters, i.e. the lengths of their sides, while the dimensionless parameters, the angles at the vertices, are identical for the two triangles. In other words, for two observers having appropriately different length scales their similar triangles are identical. Analogously, *physical phenomena are called similar if they differ only in respect of the numerical values of the dimensional governing parameters, the values of the corresponding dimensionless parameters* $\Pi_1, \ldots, \Pi_{n-k}$ *being identical*. In other words, for two observers having appropriately different basic units of measurements, (physically) similar phenomena are identical.

In accordance with the definition we have just adopted for similar phenomena, the quantities $\Pi_1, \ldots, \Pi_{n-k}$ are called the *parameters of similitude*.

Assume that we propose to model a certain phenomenon. We call this the *prototype*. We require that the *model* we wish to design has the desired properties of the prototype and that it is a phenomenon *similar* to the prototype. Therefore we have the following relationship between a quantity a to be derived and its governing parameters $a_1, \ldots, a_n$:

$$a = f(a_1, \ldots, a_k, a_{k+1}, \ldots, a_n) . \tag{2.22}$$

The function f is the same for both the prototype and the model, even though the numerical values of the governing parameters $a_1, \ldots, a_n$ and the derived quantity may differ. Thus the relation (2.22) for the prototype takes the form

$$a^{(P)} = f(a_1^{(P)}, \ldots, a_k^{(P)}, a_{k+1}^{(P)}, \ldots, a_n^{(P)}) . \tag{2.23}$$

From now on the label P will be used to refer to properties of the prototype. For the model, the relation (2.22) is similar in form but the numerical values of the governing and derived parameters (those yet to be determined) are different:

$$a^{(M)} = f(a_1^{(M)}, \ldots, a_k^{(M)}, a_{k+1}^{(M)}, \ldots, a_n^{(M)}) . \tag{2.24}$$

The label M is used to refer to properties of the model. Via dimensional analysis we obtain

$$\begin{aligned} \Pi^{(P)} &= \Phi(\Pi_1^{(P)}, \ldots, \Pi_{n-k}^{(P)}) , \\ \Pi^{(M)} &= \Phi(\Pi_1^{(M)}, \ldots, \Pi_{n-k}^{(M)}) , \end{aligned} \tag{2.25}$$

where the function Φ is the same in both cases, since (see the preceding section) it can be expressed in terms of the function f in the same way in each case. The quantities $\Pi^{(P)}, \Pi^{(M)}, \Pi_1^{(P)}, \Pi_1^{(M)}, \ldots$ are dimensionless parameters constructed according to (2.19).

2.3.2 Scaling from the model to the prototype

Since the model and prototype phenomena are physically similar, the following conditions must be satisfied according to the definition of similar phenomena given above:

$$\Pi_1^{(M)} = \Pi_1^{(P)}, \qquad \ldots, \qquad \Pi_{n-k}^{(M)} = \Pi_{n-k}^{(P)} . \tag{2.26}$$

Conditions (2.26) are called the criteria of similitude.

Hence, as stated above:

$$\Phi(\Pi_1^{(M)}, \ldots, \Pi_{n-k}^{(M)}) = \Phi(\Pi_1^{(P)}, \ldots, \Pi_{n-k}^{(P)}) \tag{2.27}$$

and, in accordance with (2.25) and indeed with our starting point, the dimensionless parameters to be determined for the model and for the prototype are equal:

$$\Pi^{(P)} = \Pi^{(M)} . \tag{2.28}$$

Returning to the dimensional parameters $a, a_1, \ldots, a_k$, using (2.19) we find that

$$a^{(P)} = a^{(M)} \left(\frac{a_1^{(P)}}{a_1^{(M)}} \right)^p \cdots \left(\frac{a_k^{(P)}}{a_k^{(M)}} \right)^r . \tag{2.29}$$

Equation (2.29) constitutes a simple rule for scaling the results of measurements from a model to the prototype. It was precisely in order to use this relation that it was necessary to require that the model be similar to the prototype.

2.3.3 Final choice of the parameters of the model

The model parameters $a_1^{(M)}, \ldots, a_k^{(M)}$ may be selected arbitrarily, keeping in mind maximum simplicity and convenience in modeling. The conditions for similarity between the model and the prototype, i.e. the equality (2.26) of the parameters of similitude $\Pi_1, \ldots, \Pi_{n-k}$ for both the model and the prototype, show how the remaining governing parameters $a_{k+1}^{(M)}, \ldots, a_n^{(M)}$ must be chosen in order to maintain this similarity. These conditions are as follows.

$$\text{For} \quad \Pi_1^{(M)} = \Pi_1^{(P)}, \qquad a_{k+1}^{(M)} = a_{k+1}^{(P)} \left(\frac{a_1^{(M)}}{a_1^{(P)}} \right)^{p_{k+1}} \cdots \left(\frac{a_k^{(M)}}{a_k^{(P)}} \right)^{r_{k+1}} .$$

$$\vdots \qquad\qquad \vdots \tag{2.30}$$

$$\text{For} \quad \Pi_{n-k}^{(M)} = \Pi_{n-k}^{(P)}, \qquad a_n^{(M)} = a_n^{(P)} \left(\frac{a_1^{(M)}}{a_1^{(P)}} \right)^{p_n} \cdots \left(\frac{a_k^{(M)}}{a_k^{(P)}} \right)^{r_n} .$$

   The simple definitions and statements presented above describe the entire content of the theory of physical similitude: we emphasize there is nothing more in this theory. The examples presented below will demonstrate how to use this theory.

2.4 Examples. Classical parameters of similitude

Here we demonstrate how to use the theory of similitude discussed in the previous section. Along the way the reader will become familiar with the classical similitude parameters often encountered in the literature.

2.4.1 Steady motion of a body in a fluid filling a very large vessel. Reynolds parameter

We make two assumptions. First, because of the very large size of the vessel, we may assume that any effects related to the boundaries of the vessel can be neglected. The second assumption is that the velocity of fluid motion is small in comparison with the velocity of sound (cf. Section 2.4.3 below), so that pressure variations due to the motion of the body are small enough to allow the fluid density ρ to be considered constant.

   The first requirement regarding a model's body and its motion is the following: the body of the model should be geometrically similar to that of the prototype, and the direction of the velocity vector with respect to the principal axes of the model's body should be the same as that of the prototype. This requirement follows from the equality of the geometric parameters of similitude and the kinematic parameters of similitude for the model and the prototype.

   To establish the dynamic parameters of similitude we note that the dimensional governing parameters of the motion under consideration are the characteristic length scale of the body, for instance the diameter of its maximum cross-section D, the magnitude of the velocity U, the density of the fluid ρ and its viscosity η. Their dimensions are as follows:

$$[D] = L, \quad [U] = LT^{-1}, \quad [\rho] = ML^{-3}, \quad [\eta] = ML^{-1}T^{-1} . \qquad (2.31)$$

Clearly, in this case $n = 4$ and $k = 3$, so here there is only one dynamic parameter of similitude,

$$\Pi_1 = \frac{\rho\, UD}{\eta} . \qquad (2.32)$$

   As proposed by A. Sommerfeld, this parameter is called the Reynolds parameter or Reynolds number, in honour of the English scientist and engineer Osborne Reynolds, who was one of the first to apply ideas of similitude to hydrodynamics. The conventional symbol for this parameter is Re.

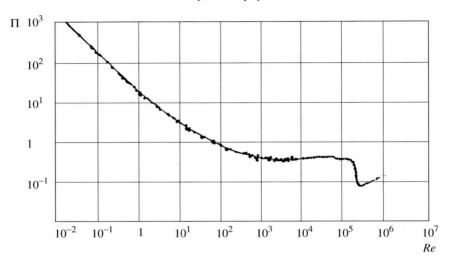

Figure 2.7 Dimensionless drag force on a sphere as a function of the Reynolds number. The data from various experiments shown here turn out to lie on a single curve, which indicates that the Reynolds number is the only parameter governing the global structure of the flow. The complicated nature of the curve indicates that the flow regime changes with Reynolds number. From Barenblatt (2003).

The most important quantity to be obtained from modeling is the drag force on the body, $\mathcal{F}$. Its dimension is MLT^{-2}, which obviously can be expressed, via the dimensions of the governing parameters ρ, U and D having independent dimensions, as $[\mathcal{F}] = [\rho][U]^2[D]^2$. Therefore the corresponding dimensionless parameter can be introduced as

$$\Pi = \frac{\mathcal{F}}{\frac{1}{2}\rho\, U^2 S} \sim \frac{\mathcal{F}}{\rho\, U^2 D^2} \tag{2.33}$$

where $S \sim D^2$ is the maximum cross-sectional area of the body, introduced according to engineering tradition, as is the factor $\frac{1}{2}$.

The function $\Pi(Re)$ for flow past a sphere is shown in Figure 2.7. The experimental data are not shown in detail; it is enough to say that, to high accuracy, data from a large number of different experiments performed by different workers lie on a single curve. As we see, the curve seems to be rather complicated: regions in which $\Pi(Re)$ varies smoothly alternate with sharp falls and rises, and there are also regions where Π is almost Re-independent. This all indicates that flow regimes vary with Reynolds number, which in this case is the only parameter that governs the structure of flow as a whole.

The motion of a model is usually implemented in the same fluid as that in which the prototype moves. The similitude condition that the Reynolds parameter (2.32) should be the same for both model and prototype motion requires in this case that the product UD should be identical for model and prototype. Therefore the velocity at which the dynamically similar model should move is inversely proportional to the scaling. According to this and to relation (2.33), the drag forces for the prototype and the dynamically similar model are identical and the scaling coefficient for the drag force is equal to unity in this case.

2.4.2 Motion of a streamlined boat at high speed. The Froude parameter

The governing parameters in this case will be as follows: the characteristic length ℓ of the boat, the gravitational acceleration g, the fluid density ρ and the speed U of the boat. We neglect the contribution of viscosity, because the drag on the boat for rapid motion, in which we are primarily interested, is due to surface waves created by the boat; therefore the gravitational acceleration is included in the list of governing parameters. The dimensions of the governing parameters are

$$[\ell] = L, \quad [g] = LT^{-2}, \quad [\rho] = ML^{-3}, \quad [U] = LT^{-1}. \tag{2.34}$$

Therefore $n = 4$ and $k = 3$. So, in addition to the geometric and kinematic similarity parameters, there is again only one dynamic parameter of similitude:

$$\Pi_1 = \frac{U}{\ell^{1/2}g^{1/2}}. \tag{2.35}$$

This parameter is called the Froude number, or Froude parameter (the conventional symbol is Fr) in honour of the English shipbuilder William Froude.

The dimension of the drag force $\mathcal{F}$, in the same class L, M, T, is $[\mathcal{F}] = LMT^{-2}$, so that $[\mathcal{F}] = [\rho][g][\ell]^3$. Therefore the law for the scaling of the drag force from the model up to the prototype is

$$\mathcal{F}^{(P)} = \mathcal{F}^{(M)} \left(\frac{\ell^{(P)}}{\ell^{(M)}} \right)^3. \tag{2.36}$$

Thus the drag force is proportional to the cube of the modeling scale. Naturally, we have assumed that g, the acceleration due to gravity, and the fluid density ρ are the same for the model and prototype; any correction of formula (2.36), if necessary, should be obvious.

In this example we have neglected the role of viscosity. If we had not, a second dynamical parameter of similitude would have appeared (see the previous example), namely the Reynolds number $Re = \rho U\ell/\eta$. Modeling in which both similarity parameters, the Froude number and the Reynolds number, are taken into account turns out to be impossible in a given fluid. Indeed, to do this the products $U\ell$ and $U/\ell^{1/2}$ would have to be identical for the model and the prototype, and this is

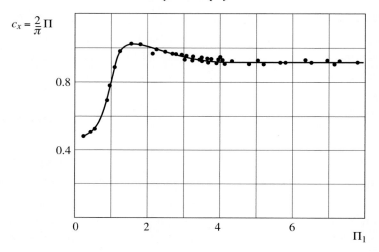

Figure 2.8 The dimensionless drag on a sphere, Π (times $2/\pi$), as a function of the dimensionless governing parameter $\Pi_1 = U/c$, the Mach number (Chernyi, 1961). The quantity Π approaches a constant for large values of Π_1.

possible only when modeling at full scale, which is pointless. Therefore, for our illustration we have restricted ourselves to the idealized case of a streamlined boat in rapid motion since in this case the viscous contribution to the drag force is small compared with the wave drag. In reality, in ship-building practice the viscous drag contribution is taken into account using specially developed techniques.

2.4.3 Steady motion of a body in a gas at high velocity. The Mach parameter

At high velocity the drag on the motion of a body is mainly due to the inertia of the gas being displaced by the body, and we can neglect the influence of viscosity. The governing dimensional parameters are the characteristic length size of the body, e.g. the diameter of its maximum cross-section D, the magnitude of the body's velocity U and the fluid's properties, its density ρ, the speed of sound c and the adiabatic index γ. As in the example in Section 2.4.1, we will consider the volume occupied by the gas to be infinite so that we can neglect edge phenomena, for instance those related to the reflections of compression–rarefaction waves from boundaries. Therefore the governing parameters and their dimensions are as follows:

$$[D] = L, \quad [U] = LT^{-1}, \quad [\rho] = ML^{-3}, \quad [c] = LT^{-1}, \quad [\gamma] = 1. \tag{2.37}$$

We see that for this case also, in addition to the geometric and kinematic parameters only one dynamic dimensionless parameter of similitude appears,

$$\Pi_1 = \frac{U}{c} . \tag{2.38}$$

This parameter is named the Mach number in honour of the Austrian natural philosopher Ernst Mach who performed pioneering experiments with shock waves in a gas; the conventional symbol is M. The dimensionless drag force is introduced in exactly the same way as in (2.33), so that in this case

$$\Pi = \frac{\mathcal{F}}{\rho\, U^2 D^2} = \Pi(M, \gamma) . \tag{2.39}$$

In Figure 2.8 the function $\Pi(M, \gamma)$ is shown for flow around a sphere. As in the first example, the Mach number M is the only parameter which governs the structure of the flow as a whole. It is interesting to note that the function $\Pi(M, \gamma)$ is close to a constant at large M, so the formula for the drag assumes the very simple form

$$\mathcal{F} = \text{const} \times \rho U^2 D^2, \tag{2.40}$$

and the influence of the speed of sound disappears.

3

The ideal incompressible fluid approximation: general concepts and relations

3.1 The fundamental idealization (model). Euler equations

Using the continuum approach formulated above and the Euler representation, i.e. following the events at fixed points of the observer's reference system, we arrive at mass and momentum balance equations, which can be represented in a unified form as

$$\partial_t \rho + \nabla \cdot \mathbf{j} = 0 \qquad (3.1)$$

and

$$\partial_t \rho \mathbf{u} + \nabla \cdot \mathbf{\Pi} = \rho \mathbf{F} . \qquad (3.2)$$

Here $\mathbf{j} = \rho \mathbf{u}$ is the flux density vector, with components $j_i = \rho u_i$, and $\mathbf{\Pi}$ is the second-rank tensor

$$\mathbf{\Pi} = \rho \mathbf{u} \otimes \mathbf{u} - \boldsymbol{\sigma},$$

having components

$$\Pi_{ij} = \rho u_i u_j - \sigma_{ij} ,$$

which is known as the momentum flux density tensor.

As we noticed before, the system (3.1), (3.2) is not closed: it contains four equations and 13 unknowns. As we saw in Chapter 1, the law of mass balance led us to a single equation with four unknowns: the addition of the momentum balance equation only increased the deficit. It is clear that using additional balance laws would increase, rather than decrease, the deficit of equations when considering the general motion of the continuum. Therefore we will take a different path, narrowing the class of motions under consideration.

With this in mind, represent the stress tensor $\boldsymbol{\sigma}$ in the form

$$\boldsymbol{\sigma} = -p\mathbf{I} + \boldsymbol{\tau} , \qquad (3.3)$$

and assume that we have a certain procedure for independently measuring p. Generally speaking, we do not assume that the scalar p is equal to $-\operatorname{trace}\boldsymbol{\sigma}/3$, where $\operatorname{trace}\boldsymbol{\sigma}$, the trace of the stress tensor, is the linear invariant of this tensor equal to the sum of the diagonal components: $\operatorname{trace}\boldsymbol{\sigma} = \sigma_{\alpha\alpha}$. Therefore, generally speaking, $\operatorname{trace}\boldsymbol{\tau}$ is different from zero. The scalar p stands for pressure.

Changing notation, equations (3.1), (3.2) can be rewritten as

$$\frac{1}{\rho}\frac{d\rho}{dt} + \nabla \cdot \mathbf{u} = 0 ,$$

$$\frac{d\mathbf{u}}{dt} = -\frac{1}{\rho}\nabla p + \mathbf{F} + \frac{1}{\rho}\nabla \cdot \boldsymbol{\tau} . \tag{3.4}$$

We will now consider a special class of continuum motions such that the following strong inequalities hold at all times in the entire space occupied by the continuum:

(1) $|\nabla \cdot \boldsymbol{\tau}| \ll |\nabla p|$;
(2) *The order of magnitude of the quantity $\left|\frac{1}{\rho}\frac{d\rho}{dt}\right|$ is much less than the order of magnitude of the largest terms of the sum $\nabla \cdot \mathbf{u} = \partial_\alpha u_\alpha = \partial_1 u_1 + \partial_2 u_2 + \partial_3 u_3$.*

For such motions it is possible to neglect in the first equation of (3.4) the term involving the variation of density with time (the first term on the left-hand side) and in the second equation of (3.4) the term on the right-hand side representing spatial variations of the stress tensor $\boldsymbol{\tau}$. Assuming further that the medium is homogeneous (this special assumption is independent and for many important classes of motions is not valid, even if (1) and (2) are valid), it follows that the density ρ is a given constant quantity. Thus, neglecting the two terms in equations (3.4) involving variations of ρ or $\boldsymbol{\tau}$, we obtain a closed system of equations valid for motions in the class under consideration:

$$\partial_t \mathbf{u} + (\mathbf{u} \cdot \nabla) \cdot \mathbf{u} = -\frac{1}{\rho}\nabla p + \mathbf{F}, \tag{3.5}$$

$$\nabla \cdot \mathbf{u} = 0 . \tag{3.6}$$

The model constructed in this way is called the *ideal incompressible fluid approximation*, and the imaginary idealized continuous medium for which these equations would be exactly valid is called an *ideal incompressible fluid*.

We stress that, for the class of motions corresponding to the approximation of an ideal incompressible fluid, the system of equations (3.5), (3.6) contains four unknowns for four scalar equations: the scalar pressure p and the three components of the velocity vector $\mathbf{u}$. Thus the system is closed.

We emphasize two points. The first one is related to the approximation (2). It is impossible to replace (2) by the assumption that the spatial density variation $\delta\rho$ is small, i.e.

$$\left|\frac{\delta\rho}{\rho}\right| \ll 1 \,.$$

This assumption is not sufficient. For instance, when studying sound propagation (see Chapter 10), the strong inequality $\left|\frac{\delta\rho}{\rho}\right| \ll 1$ holds, but nevertheless it would be incorrect to neglect the term $\frac{1}{\rho}\frac{d\rho}{dt}$ in (3.4) because it is of the same magnitude as terms in the sum $\partial_\alpha u_\alpha$. The term $\frac{1}{\rho}\frac{d\rho}{dt}$ is exactly equal to $-\partial_\alpha u_\alpha$, but the approximation we are dealing with is valid in cases when the order of magnitude of the sum $\partial_\alpha u_\alpha$ is much less than the order of magnitude of the leading terms of the sum. (Just to clarify: let $\frac{1}{\rho}\frac{d\rho}{dt}$ be, in some units, equal to 1 and the terms of the sum be equal to -1000, 499 and 500.)

Furthermore, the character of assumptions (1) and (2) should be emphasized. Using the approximation of an ideal incompressible fluid we can consider only motions satisfying assumptions (1) and (2). We should not be mesmerized by the term "ideal fluid", as sometimes happens. When the ideal incompressible fluid approximation is used to model the real motions of real bodies, the researcher should pay special attention to clarifying whether this approximation is valid for the motions under consideration, independently of whether fluid motion or the motion of a solid body is considered. (See especially the instructive first example in the next chapter.)

For instance, in Chapter 40 of his excellent course in physics (Feynman, 2006), Vol. II, pp. 40–3, Professor R. Feynman writes:

> First, we will discuss fluid motions in a purely abstract, theoretical way, and then consider special examples. To describe the motion of a fluid we must give its properties at every point. For example, at different places, the water (let us call the fluid "water") is moving with different velocities.
>
> For this chapter we are going to suppose that the liquid is "thin" in the sense that the viscosity is unimportant, so we will omit $\mathbf{f}_{\mathrm{visc}}$.[1] When we drop the viscosity term, we will be making an approximation which describes some ideal stuff rather than real water. John von Neumann was well aware of the tremendous difference between what happens when you don't have the viscous terms and when you do, and he was also aware that, during most of the development of hydrodynamics until about 1900, almost the main interest was in solving beautiful *mathematical* problems with this approximation which had almost nothing to do with real fluids. He characterized the theorist who made such analyses as a man who studied "dry water". Such analyses leave out an *essential* property of the fluid. It is because we are leaving this property out of our calculations in this chapter that we have given it the title "The flow of dry water". We are postponing a discussion of *real* water to the next chapter.

---

[1] In the Feynman lectures, $\mathbf{f}_{\mathrm{visc}}$ is equivalent to $\nabla\tau$ in the second equation in (3.4).

If the reader thinks that s/he will find here the *real water* discussed in the next chapter of Feynman's course, s/he will be disappointed: it is yet another model, another idealization, which we will discuss later.

Feynman should not reproach researchers for finding solutions to problems that are not similar to certain motions of real "wet" fluids. Being mesmerized by the term "ideal fluid", researchers want to find a correspondence between such solutions and the observed properties of real flows of real fluids. Very often there exists such a correspondence but not where it is sought. We will see later that the ideal incompressible fluid approximation corresponds perfectly to some motions of soil or air (and also sometimes those of tank armor), but not necessarily those of water. Therefore we will not hesitate to use the nickname "dry fluid" and to study the motions of this idealized continuum. However, we must always remember that in comparing the results of such studies with the real motions of real bodies we must pay attention to the suitability of the assumed idealization.

In the system of orthogonal Cartesian coordinates x_i ($i = 1, 2, 3$) the equations of motions of continuum in the ideal incompressible fluid approximation (3.5), (3.6) assume the form

$$\partial_t u_i + u_\alpha \partial_\alpha u_i = -\frac{1}{\rho} \partial_i p + F_i \,, \tag{3.7}$$

$$\partial_\alpha u_\alpha = 0 \,. \tag{3.8}$$

Equations (3.7) are called the Euler equations. The system (3.7), (3.8) should be complemented by an initial condition imposed on the velocity (but not on the pressure, because the time derivative of the pressure does not enter the system) and by appropriate boundary conditions, which will be discussed later.

3.2 Decomposition of the velocity field in the vicinity of an arbitrary point. The vorticity. The strain-rate tensor

Consider the velocity distribution in a small vicinity of an arbitrary point **x**. We remain within the framework of the Eulerian approach and the assumption that the velocity field is sufficiently smooth.[2] Expanding the velocity field in a Taylor series and restricting ourselves to first-order quantities, we obtain

$$\mathbf{u}(\mathbf{x} + \mathbf{r}) = \mathbf{u}(\mathbf{x}) + (\text{grad}\,\mathbf{u}) \cdot \mathbf{r} \,. \tag{3.9}$$

Here **r** is the radius vector of a point in the vicinity of the point **x**; its length scale is assumed to be small (in comparison with the characteristic length scale of the flow under consideration). The second-rank tensor grad **u** has components $\partial_j u_i$ in

---

[2] In fact, this assumption is non-trivial, and should be carefully investigated in any particular case.

the orthonormal Cartesian coordinate system. The tensor grad **u** can be represented as the sum of an antisymmetric tensor **W** and a symmetric tensor **D**:

$$\text{grad } \mathbf{u} = \mathbf{W} + \mathbf{D}, \tag{3.10}$$

where

$$\mathbf{W} = \frac{1}{2}(\text{grad } \mathbf{u} - \text{grad } \mathbf{u}^T), \qquad \mathbf{D} = \frac{1}{2}(\text{grad } \mathbf{u} + \text{grad } \mathbf{u}^T),$$

or, in components,

$$W_{ij} = \frac{1}{2}(\partial_j u_i - \partial_i u_j), \qquad D_{ij} = \frac{1}{2}(\partial_i u_j + \partial_j u_i).$$

The superscript T denotes the transposed tensor, with components $\partial_i u_j$. The tensor **D** is called the strain-rate tensor. Note that **D** is a quantity of dimension T^{-1}. We will discuss this tensor in more detail later, when we consider the model of a Newtonian viscous fluid. We note here only that if the whole motion of the continuum in the neighborhood under consideration is reduced to a displacement like that of a solid body, the strain-rate tensor vanishes.

Furthermore, the tensor **W** is antisymmetric; therefore the vector **W** · **r** is equal to a vector product:

$$\mathbf{W} \cdot \mathbf{r} = [\boldsymbol{\omega}, \mathbf{r}]. \tag{3.11}$$

Here, $\boldsymbol{\omega}$ is the *axial vector* of the antisymmetric tensor **W**, equal to

$$\boldsymbol{\omega} = \frac{1}{2} \text{ curl } \mathbf{u}. \tag{3.12}$$

If the reader does not recall these relations, as well as the subsequent relations (3.13), and (3.16), s/he can verify them very easily in component form.

Thus, after substitution in (3.9), we find that *the velocity in the neighborhood of a given point can be represented as the sum of the translation velocity of the center of the neighborhood, the velocity of solid body rotation around the center with angular velocity $\boldsymbol{\omega}$ and the velocity of deformation.* This fundamental statement is called *von Helmholtz' theorem.* Therefore the local angular velocity of rotation of the continuum is determined by the vorticity vector $\boldsymbol{\omega}$, and this makes this vector a very important property of the motion. It is expressed (see (3.12)) via the velocity field and it is therefore possible to derive an equation, on the basis of Euler's equations, that governs the vorticity dynamics in the ideal incompressible fluid approximation. Indeed, according to a well-known formula of elementary vector analysis we have

$$(\mathbf{u} \cdot \nabla) \cdot \mathbf{u} = \text{grad } \frac{\mathbf{u}^2}{2} - [\mathbf{u}, \text{curl } \mathbf{u}]. \tag{3.13}$$

Using this formula, Euler's equation can be rewritten in a different form, called the *Lamb*[3] *representation*:

$$\partial_t \mathbf{u} + \text{grad}\,\frac{\mathbf{u}^2}{2} - [\mathbf{u}, \text{curl}\,\mathbf{u}] = \mathbf{F} - \frac{1}{\rho}\,\text{grad}\,p\,. \qquad (3.14)$$

The next step is the application of the curl operation to both sides of (3.14). Bearing in mind (3.12) and that $\text{curl}\,\text{grad}\,f \equiv 0$ for an arbitrary scalar function f, we obtain

$$\partial_t \boldsymbol{\omega} = \text{curl}[\mathbf{u}, \boldsymbol{\omega}] + \frac{1}{2}\,\text{curl}\,\mathbf{F}\,. \qquad (3.15)$$

As is well known, for two arbitrary vectors $\mathbf{a}$ and $\mathbf{b}$ the identity

$$\text{curl}[\mathbf{a} \cdot \mathbf{b}] = (\mathbf{b} \cdot \nabla)\mathbf{a} - (\mathbf{a} \cdot \nabla)\mathbf{b} + \mathbf{a}\,\text{div}\,\mathbf{b} - \mathbf{b}\,\text{div}\,\mathbf{a} \qquad (3.16)$$

holds, so that using the relation $\text{div}\,\text{curl}\,\mathbf{u} = 0$, valid for an arbitrary vector $\mathbf{u}$, and the continuity equation for an ideal incompressible fluid (3.6), $\text{div}\,\mathbf{u} = 0$, we come to the basic equation controlling the dynamics of the vorticity field,

$$\partial_t \boldsymbol{\omega} = (\boldsymbol{\omega} \cdot \nabla)\mathbf{u} - (\mathbf{u} \cdot \nabla)\boldsymbol{\omega} + \frac{1}{2}\,\text{curl}\,\mathbf{F} \qquad (3.17)$$

or

$$\frac{d\boldsymbol{\omega}}{dt} = (\boldsymbol{\omega} \cdot \nabla)\mathbf{u} + \frac{1}{2}\,\text{curl}\,\mathbf{F}\,. \qquad (3.18)$$

This equation shows that the rate of vorticity formation for a given particle (the derivative $d\omega/dt$, not $\partial_t \omega$) is determined by two additive factors. The first is the vorticity production by the existing vorticity due to the non-uniformity of the velocity field, and the second is the vorticity production by the vortex part of the mass force field $\mathbf{F}$. If the latter is potential, i.e. if there exists a force potential V such that $\mathbf{F} = -\text{grad}\,V$, the second term on the right-hand side of (3.18) disappears and this equation takes the form

$$\frac{d\boldsymbol{\omega}}{dt} = (\boldsymbol{\omega} \cdot \nabla)\mathbf{u}\,. \qquad (3.19)$$

3.3 Irrotational motions. Lagrange's theorem. Potential flows

The motions of a continuous medium are called *irrotational* if in the whole region occupied by the moving continuous medium the vorticity vector $\boldsymbol{\omega}$ is identically equal to zero.

The ideal incompressible fluid approximation is distinguished by the following fundamental result.

---

[3] Horace Lamb was an outstanding British fluid dynamicist.

Theorem (Lagrange's theorem)
(1) *If the system of mass forces acting on the continuous medium is a potential
    one, i.e. if there exists a scalar function V such that* $\mathbf{F} = -\operatorname{grad} V$*, and*
(2) *if at the initial time instant* $t = t_0$ *the motion is irrotational,*

then, in the ideal incompressible fluid approximation, it remains irrotational at
$t > t_0$.

Indeed, due to (1), $\operatorname{curl}\mathbf{F} = 0$, so that the equation for the dynamics of vorticity (3.18) takes the form (3.19). The velocity $\mathbf{u}$ is a smooth function: therefore the coefficients of this equation are also smooth functions. Hence the trivial solution $\boldsymbol{\omega} \equiv 0$ satisfying the condition $\boldsymbol{\omega} = 0$ at $t = t_0$ is unique, and in the ideal incompressible fluid approximation the vorticity $\boldsymbol{\omega}$ is identically equal to zero in the entire region of the motion, also at $t > t_0$.

Let us present the proof of this important statement in more detail. Multiplying (3.19) by $2\boldsymbol{\omega}$, we obtain

$$\partial_t \omega^2 + (\mathbf{u} \cdot \nabla)\omega^2 = 2\boldsymbol{\omega}(\boldsymbol{\omega} \cdot \nabla) \cdot \mathbf{u}, \tag{3.20}$$

or, in coordinates,

$$\frac{d\omega^2}{dt} = 2\omega_\alpha(\omega_\beta \partial_\beta u_\alpha) = 2\omega^2 |\partial_\beta u_\alpha| \frac{\omega_\alpha}{|\omega|} \frac{\omega_\beta}{|\omega|} \frac{\partial_\beta u_\alpha}{|\partial_\beta u_\alpha|}. \tag{3.21}$$

However,

$$\left| \frac{\omega_\alpha}{|\omega|} \frac{\omega_\beta}{|\omega|} \frac{\partial_\beta u_\alpha}{|\partial_\beta u_\alpha|} \right| = C_{\alpha\beta} < 1,$$

so that

$$\frac{d\omega^2}{dt} = 2\omega^2 |\partial_\beta u_\alpha| C_{\alpha\beta} \le \omega^2 K, \tag{3.22}$$

where K is finite owing to the smoothness of the velocity field $\mathbf{u}(\mathbf{x})$, and so Lagrange's theorem is proved.

In particular, all motions which start from a state of rest, i.e. for which $\mathbf{u} \equiv 0$ at $t = t_0$ so that $\boldsymbol{\omega} \equiv 0$ at $t = t_0$, remain irrotational in the ideal incompressible fluid approximation if the mass force is potential.

The irrotational character of the motion introduces essential mathematical simplifications to its analysis. Indeed, if $\operatorname{curl}\mathbf{u} = 0$ then, as it is easy to show, there exists a scalar function ϕ such that

$$\mathbf{u} = \operatorname{grad}\phi. \tag{3.23}$$

Such motions are called *potential*, and the function ϕ is called the *velocity potential*. Putting (3.23) into the continuity equation $\operatorname{div}\mathbf{u} = 0$, we obtain that the velocity potential is a harmonic function satisfying the *Laplace equation*

$$\Delta\phi = 0 . \tag{3.24}$$

We emphasize that although we did not use the Euler equations explicitly to derive (3.24), the very fact that the flow was potential, as follows from previous arguments, was essentially related to these equations. A natural question arises, whether an ideal incompressible fluid approximation is a *necessary* condition for the flow to be potential. In fact, it is not so: flows need be neither ideal nor incompressible to be potential. See, in particular, the instructive book of Joseph *et al.* (2008).

The Laplace equation (3.24) has been much investigated, and many well-posed boundary value problems are known for this equation, leading to the existence and uniqueness of the solution. The best known ones are the Dirichlet and Neumann problems. In the first of these problems, the values of the potential are prescribed on the boundary. We do not know yet the physical meaning of the potential, and therefore we postpone consideration of the Dirichlet problem until the next chapter.

In the Neumann problem the derivative of the potential along the normal **n** to the boundary, $\partial\phi/\partial n$, is prescribed on the boundary. This condition has a clear physical sense. It corresponds to the flow of a continuous medium around a body when the flow is not separated from the body surface. Indeed, at the boundary of a solid body around which the continuous medium moves, a condition of non-penetration and non-separation should be fulfilled so that the normal components of velocity of the body and medium coincide. From this, Neumann's condition follows:

$$u_n = \frac{\partial\phi}{\partial n} = v_n , \tag{3.25}$$

where u_n is the velocity component of the medium normal to the body surface, and v_n is the projection of the velocity of the body onto the normal to its surface. In particular, if the body is at rest then $u_n = 0$.

If the length scale of the region occupied by the continuous medium is much larger than the size of the body, it is sometimes appropriate to consider this region as infinite. In this case the condition at infinity should be prescribed. Often it is the velocity at infinity that is given, so that

$$\text{grad}\,\phi \to \mathbf{U} , \qquad \mathbf{x} \to \infty. \tag{3.26}$$

The boundary conditions (3.25) and (3.26), and sometimes some additional conditions (see below), uniquely determine the function ϕ up to an immaterial additive constant, and so uniquely determine the velocity field $\mathbf{u} = \text{grad}\,\phi$. It is important that in the ideal incompressible fluid approximation the flowing medium *does not stick to the boundary of the body but slips along it.* Indeed, the tangential projection of the velocity, $u_s = \partial \phi / \partial s$, where s is the direction tangential to the boundary, is, generally speaking, different from v_s, the projection of the velocity of the body tangential to the boundary. This means that problems where the tangential component of the velocity at the boundary is influenced by some interaction with the body surface cannot be considered in the ideal incompressible fluid approximation. An example of such a problem is the motion around a body with a non-slip condition. We will see this in more detail later when we consider the viscous incompressible fluid approximation.

3.4 Lagrange–Cauchy integral. Bernoulli integral

After solving the Laplace equation for the velocity potential under certain appropriate conditions we obtain the velocity potential ϕ and the velocity itself, $\mathbf{u} = \text{grad } \phi$. We turn now to determining the pressure. Substituting $\mathbf{u} = \text{grad } \phi$ into the Euler equation (in the Lamb representation) and assuming the mass force also to be a potential one, so that $\mathbf{F} = - \text{grad } V$, we reduce the Euler equation to the form

$$\partial_t \text{ grad } \phi + \text{grad } \frac{\mathbf{u}^2}{2} + \text{grad } V + \frac{1}{\rho} \text{ grad } p = \text{grad} \left(\partial_t \phi + \frac{\mathbf{u}^2}{2} + V + \frac{p}{\rho} \right) = 0. \quad (3.27)$$

The *Lagrange–Cauchy integral* is obtained by integration:

$$\partial_t \phi + \frac{1}{2} (\text{grad } \phi)^2 + V + \frac{p - p_0}{\rho} = 0. \quad (3.28)$$

Here p_0 is the pressure at a certain point and depends, generally speaking, on time. Note that the potential ϕ is defined up to an arbitrary additive function of time $-f(t)$: the velocity field remains the same if we pass from ϕ to the potential ϕ', where

$$\phi' = \phi + \int f(t) \, dt , \qquad \text{grad } \phi' = \text{grad } \phi = \mathbf{u} . \quad (3.29)$$

Therefore we can add to the right-hand side of the Lagrange–Cauchy integral (3.28) an arbitrary function of time, redefining the velocity potential as necessary.

   The Lagrange–Cauchy integral should be distinguished from the *Bernoulli integral*, although in some special cases these integrals coincide. To derive the Bernoulli integral assume that:

(1) The motion of the continuous medium can be considered in the approximation of an *ideal fluid* (incompressibility is not necessary).
(2) The system of mass forces $\mathbf{F}$ acting on the continuous medium is a *potential one*: $\mathbf{F} = - \text{grad } V$.
(3) The motion is *steady*, i.e. the velocity field does not depend on time.

We do not assume the motion to be potential and, generally speaking, it can be rotational.

As a preliminary we introduce an important concept. Consider a vector field $\mathbf{a}(\mathbf{x}, t)$. The lines which are tangent to the vector $\mathbf{a}$ at every point are called the *vector lines* of the field $\mathbf{a}(\mathbf{x}, t)$. Evidently, the vector lines satisfy the equation (recall the notation $[\mathbf{a}, \mathbf{b}]$ for the vector product of two vectors $\mathbf{a}$ and $\mathbf{b}$)

$$[d\mathbf{x}, \mathbf{a}] = 0 \tag{3.30}$$

at every point. So, in a system of orthonormal Cartesian coordinates x_1, x_2, x_3, the following differential relations hold:

$$\frac{dx_1}{a_1} = \frac{dx_2}{a_2} = \frac{dx_3}{a_3}. \tag{3.31}$$

Here a_1, a_2, a_3 are the x_1, x_2, x_3 components of vector $\mathbf{a}$.

Two special families of vector lines are important for our purpose: the family for which $\mathbf{a}(\mathbf{x}, t) = \mathbf{u}(x, t)$, and that for which $\mathbf{a}(\mathbf{x}, t) = \boldsymbol{\omega}(\mathbf{x}, t)$. In these cases the vector lines are called respectively *stream lines* and *vortex lines*.

In the case of steady motions, the stream lines coincide with trajectories of the particles of the continuum whereas, generally speaking, for unsteady motions this is not so. The equations for the stream lines have the form

$$\frac{dx_1}{u_1} = \frac{dx_2}{u_2} = \frac{dx_3}{u_3} \tag{3.32}$$

and the equations for the vortex lines are

$$\frac{dx_1}{\omega_1} = \frac{dx_2}{\omega_2} = \frac{dx_3}{\omega_3}. \tag{3.33}$$

Here ω_1, ω_2, ω_3 are the x_1, x_2, x_3 components of the vorticity vector $\boldsymbol{\omega}$. Under our assumption of steadiness the term $\partial_t \mathbf{u}$ disappears and the Euler equations in the Lamb representation take the form

$$\operatorname{grad} \frac{\mathbf{u}^2}{2} + [\operatorname{curl} \mathbf{u}, \mathbf{u}] = - \operatorname{grad} V - \frac{1}{\rho} \operatorname{grad} p . \tag{3.34}$$

Now project the vector equation (3.34) onto the stream line or vortex line passing through a given point. The vector product on the left-hand side is perpendicular both to $\mathbf{u}$ and to curl $\mathbf{u}$ so it does not contribute to the projection, which therefore in each case assumes the form

$$\frac{\partial}{\partial s} \left(\frac{\mathbf{u}^2}{2} \right) + \frac{\partial V}{\partial s} + \frac{1}{\rho} \frac{\partial p}{\partial s} = 0, \tag{3.35}$$

where s is the length reckoned along a stream line or a vortex line. The stream lines and vortex lines evidently form a two-parameter family of curves. Let us denote by

N_1, N_2 the parameters which distinguish a given curve. These could be, e.g., the coordinates of the intersection of the curve with a certain plane or surface. Along a given stream or vortex line the pressure and density depend on its length s:

$$p = p(s, N_1, N_2), \qquad \rho = \rho(s, N_1, N_2).$$

(3.36)

We remind the reader that the motion need not be incompressible; we assume, rather, that the pressure variation along the stream or vortex line is monotonic. Then we can exclude s from the two relations (3.36) and obtain

$$\rho = \rho(p, N_1, N_2), \qquad \frac{1}{\rho}\frac{\partial p}{\partial s} = \frac{\partial}{\partial s}\Phi(p, N_1, N_2), \qquad \Phi = \int \frac{dp}{\rho(p, N_1, N_2)}.$$

(3.37)

Put the last relations into (3.35); integrating, we obtain the *Bernoulli integral*

$$\frac{\mathbf{u}^2}{2} + V + \Phi(p, N_1, N_2) = C(N_1, N_2).$$

(3.38)

The derivation of the Bernoulli integral presented is due to L. I. Sedov. Note that relation (3.36), $\rho = \rho(p, N_1, N_2)$, is not a *barotropic* one when $\rho = \rho(p)$; the relation is different for different lines. Therefore the integration constant C is, generally speaking, different for different stream lines or vortex lines. For an incompressible fluid where ρ is a constant, the Bernoulli integral assumes the form

$$\frac{\mathbf{u}^2}{2} + V + \frac{p - p_0}{\rho} = C(N_1, N_2).$$

(3.39)

Evidently, for steady potential motions of an ideal incompressible fluid the Lagrange–Cauchy integral and the Bernoulli integral coincide. Indeed, because of the steadiness the term $\partial_t \phi$ in the Lagrange–Cauchy integral vanishes, and the constant in the Bernoulli integral does not depend on the stream line because the term $[\mathbf{u}, \operatorname{curl}\mathbf{u}]$ in the Euler equation vanishes identically.

3.5 Plane potential motions of an ideal incompressible fluid

The consideration of the plane potential motions of an ideal incompressible fluid is of special interest for two reasons. The first is that to study such motions we can use a very effective mathematical technique from the theory of functions of complex variables. At the same time, and this is the second reason, the study of this very special class of motions provides a useful tool in many practical and interesting real motions.

Assume that the motion under investigation obeys the ideal incompressible fluid approximation and is a potential one and that its potential depends on two space coordinates, the orthonormal Cartesian coordinates x_1, x_2, and perhaps on the time t; thus, $\phi = \phi(x_1, x_2, t)$. So the x_3 component of the velocity vanishes, and the two remaining components are identical in all planes perpendicular to the x_3 axis. Examples of such motions will be considered below.

The continuity equation for such plane motions of an incompressible fluid is written in the form

$$\partial_1 u_1 + \partial_2 u_2 = 0, \tag{3.40}$$

so that we can introduce a *stream function* $\psi(x_1, x_2)$ by the relations

$$u_1 = \partial_2 \psi, \qquad u_2 = -\partial_1 \psi \tag{3.41}$$

and use the relation (3.40) as the integrability condition, the condition for the existence of the function ψ. By the definition of a potential motion we have $\mathbf{u} = \operatorname{grad} \phi$, $u_1 = \partial_1 \phi$, $u_2 = \partial_2 \phi$. Comparing with (3.41) we obtain

$$\partial_1 \phi = \partial_2 \psi, \qquad \partial_2 \phi = -\partial_1 \psi. \tag{3.42}$$

The equations (3.42) are the well-known Cauchy–Riemann relations. They imply that the stream function ψ is a harmonic function conjugate to the potential ϕ, which is also a harmonic function, so that the function

$$w = \phi + i\psi, \tag{3.43}$$

which is called a *complex potential*, is an analytic function of a complex variable $z = x_1 + ix_2$.

This fact establishes a direct correspondence between the analytic functions of a complex variable and the plane potential flows of an ideal incompressible fluid. It means that to each analytic function there corresponds a plane potential flow and that each plane potential flow has a complex potential which is an analytic function of a complex variable.

The physical meaning of the stream function is transparent. In Figure 3.1 we show a cylindrical surface, supported by a line AB in the plane $x_1 x_2$, with generators parallel to x_3 and of unit height. Then the volume of fluid passing through this surface per unit time is equal to

$$\int_A^B (u_1 \, dx_2 - u_2 \, dx_1) = \int_A^B (\partial_1 \psi \, dx_1 + \partial_2 \psi \, dx_2) = \int_A^B d\psi = \psi(B) - \psi(A), \tag{3.44}$$

i.e. it is equal to the difference in the values of the stream function at the end and beginning of the curve AB. In particular, if the line AB is part of a stream line then, by definition, at any point on this line the tangent to it is directed along the velocity

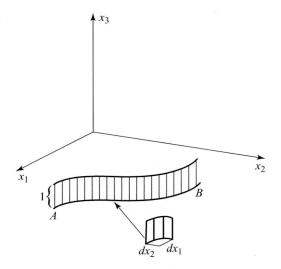

Figure 3.1 Physical meaning of the stream function.

vector at this point:

$$\frac{dx_1}{u_1} = \frac{dx_2}{u_2} \,, \qquad u_2 dx_1 - u_1 dx_2 = -d\psi = 0 \,. \tag{3.45}$$

This means that the stream function is constant along a stream line. From the elementary theory of functions of a complex variable we know that, for an analytic function $w(z)$, the families of lines of constant $\operatorname{Re} w(z)$ and constant $\operatorname{Im} w(z)$ are mutually orthogonal. Therefore the families of stream lines of constant ψ and of equipotential lines of constant ϕ form two mutually orthogonal networks of curves.

The quantity $u_1 + i u_2$ is called the *complex velocity*. According to (3.42),

$$u_1 + i u_2 = \partial_1 \phi - i \partial_1 \psi = \overline{\frac{dw}{dz}} \,, \tag{3.46}$$

where the overbar denotes a complex conjugate.

We now consider several simple examples of the plane potential flows of an ideal incompressible fluid which will be needed later.

(a) Translation flow. This flow has complex potential

$$w(z) = az \,. \tag{3.47}$$

Here $a = a_1 + i a_2$ is, generally speaking, a complex constant. By separating the real and imaginary parts of (3.47) we obtain

$$\phi = a_1 x_1 - a_2 x_2 \,, \qquad \psi = a_1 x_2 + a_2 x_1 \,, \tag{3.48}$$

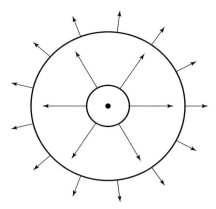

Figure 3.2 Source–sink flow.

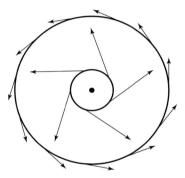

Figure 3.3 Vortex flow.

so that the velocity at all points is constant and equal to $a_1\mathbf{e}_1 - a_2\mathbf{e}_2$ (recall that $\mathbf{e}_1, \mathbf{e}_2$ are the unit vectors along the x_1 and x_2 axes).

(b) Source–sink flow. This flow has the complex potential

$$w(z) = \frac{Q}{2\pi} \ln z, \qquad \phi = \frac{Q}{2\pi} \ln r, \qquad \psi = \frac{Q}{2\pi} \theta. \tag{3.49}$$

Here r and θ are polar coordinates with origin at the point $z = 0$, so that $z = re^{i\theta}$, and Q is a real constant. From (3.49) we obtain

$$u_r = \partial_r \phi = \frac{Q}{2\pi r}, \qquad u_\theta = \frac{1}{r} \partial_\theta \phi = 0.$$

Thus the stream lines of this flow are straight lines coming from, or to, the origin, and the lines of equal potential are circles whose tangents are orthogonal to the straight stream lines (Figure 3.2). The flow corresponds to a source ($Q > 0$) or a sink ($Q < 0$) on the straight line passing through the origin in the x_3 direction. The discharge of the source–sink per unit height equals $2\pi r u_r = Q$.

(c) Vortex flow. This flow has complex potential

$$w(z) = \frac{\Gamma}{2\pi i} \ln z, \qquad \phi = \frac{\Gamma\theta}{2\pi}, \qquad \psi = \frac{\Gamma \ln r}{2\pi}, \qquad (3.50)$$

so that $u_r = 0$, $u_\theta = \Gamma/(2\pi r)$. Here, in contrast with the previous case, the stream lines are circles with the center at the origin (Figure 3.3) and orthogonal to their tangents are the equipotential lines, straight lines emerging from the origin. The *circulation* (that is, the integral $\int (\mathbf{u} \cdot d\mathbf{s})$ taken along any closed line with the origin inside) is constant and equal to Γ.

4

The ideal incompressible fluid approximation: analysis and applications

4.1 Physical meaning of the velocity potential. The Lavrentiev problem of a directed explosion

We now must clarify the direct physical meaning of the velocity potential: without understanding this it is impossible to formulate the Dirichlet boundary value problem: we have to prescribe the velocity potential at the boundary, but we do not know yet what the potential is.

Consider a body in a continuous medium which at $t = t_0$ is at rest. Assume that at $t = t_0$ each particle experiences a pressure pulse such that the pressure varies according to the law

$$p(\mathbf{x}, t) = \theta(\mathbf{x})\, \delta(t - t_0) \, . \tag{4.1}$$

Here $\theta(\mathbf{x})$ is a function of the position of the particle, and $\delta(z)$ is the generalized Dirac function. According to the simplest definition of this function, which is all we need for now,

$$\delta(z) \equiv 0 \ \text{ when } \ z \neq 0\,, \qquad \int_{-\varepsilon}^{\varepsilon} \delta(z)\, dz = 1 \, , \tag{4.2}$$

for arbitrarily small positive ε.

The motion begins from a state of rest before the pressure pulse starts. Therefore if the system of mass forces acting on the medium is a potential one, the Lagrange–Cauchy integral holds in the ideal incompressible fluid approximation:

$$\partial_t \phi + \frac{\mathbf{u}^2}{2} + V + \frac{p - p_0}{\rho} = 0 \, . \tag{4.3}$$

We put (4.1) into (4.3) and integrate from $t = t_0 - \varepsilon$ to $t = t_0 + \varepsilon$. Bearing in mind that $\phi(x, t_0 - \varepsilon)$ is identically equal to zero, because the medium before the pressure pulse was at rest, and also that $\mathbf{u}$, V and p_0 are bounded, we obtain, passing to the limit at $\varepsilon \to 0$,

$$\phi(\mathbf{x}, t_0 + 0) = -\frac{\theta(\mathbf{x})}{\rho} \ . \tag{4.4}$$

Thus, *the velocity potential is (up to a constant factor* $-1/\rho$*) the intensity of the pressure pulse needed to transfer the medium from a state of rest to the potential flow under consideration.*

   Moreover, it is not necessary to worry about creating pressure pulses at all points of the body independently. It is enough to manage the required pulses at points on the boundary of the body. The necessary pressure pulses will appear automatically in the internal points of the body: indeed, by managing the pressure pulses at the boundary we will obtain there the values of the potential ϕ needed for the potential flow under consideration. But the potential ϕ is a harmonic function – it satisfies the Laplace equation – so at $t = t_0 + 0$ it takes appropriate values at internal points of the body automatically. This determines the meaning of the Dirichlet problem for the potential motion of an ideal incompressible fluid: we prescribe at the boundary of the body the pressure pulse transforming the state of rest to a state of a given potential motion. Thus we come to an important conclusion: in the ideal incompressible fluid approximation the velocity of propagation of a disturbance is infinite.

   We are able now to consider a remarkable *direct explosion problem* originally formulated by the Soviet mathematician M. A. Lavrentiev. This mathematical problem arose from the practical one of displacing of large masses of soil. It is formulated as follows. Let us consider a mass of soil (on the left in Figure 4.1) which must be displaced to another place translationally. In fact, of necessity such a displacement was required when the Medeo canyon, near the city of Alma-Ata, the capital of the Kazakh Republic in the former Soviet Union, needed to be closed to prevent mountain torrents (known by the Turkish word "sel") from reaching the city: a serious danger existed.[1]

   Neglecting the resistance of the ambient air we see that the mass of soil needs to be transferred by a pressure pulse from its state of rest to a state of translational motion, with a velocity appropriate to displacing it in the gravity field to the desired place.[2] The required translational motion has velocity $\mathbf{U}$ with components (U_1, U_2, U_3) given by

$$U_1 = U \cos\alpha \cos\beta \ , \qquad U_2 = U \cos\alpha \sin\beta \ , \qquad U_3 = U \sin\alpha \ , \tag{4.5}$$

---

[1] The present author had a chance to watch a movie where the technical and scientific leadership of this work was attributed to Comrade D. A. Kunaev, at that time the secretary of the Kazakh Communist party. The author has different information from reliable sources.

[2] Clearly, we speak here of the basic principles for this great technical achievement. In fact, many technical details were needed to achieve the goal; they are not presented here.

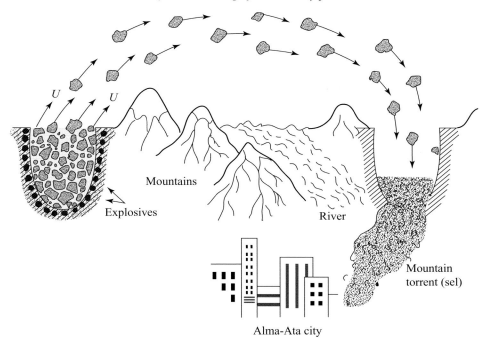

Figure 4.1 Directed explosion (Lavrentiev problem): the principles of the scheme.

where U is the magnitude of the initial velocity and α and β are the angles defining its direction. Lavrentiev proposed performing the translation by an ordinary (non-nuclear) explosion. The pressure magnitude of such an explosion is of the order of tens of thousands bars (kgf/cm^2). Lavrentiev understood that the motion of the soil in this case could be considered in the ideal incompressible fluid approximation.

Indeed, the available weakly cemented soil was able to maintain tangential shear stresses of the order at most several tens of bars. These stresses are negligibly small in comparison with the pressure which develops during the explosion. Therefore the divergence of the stress τ is negligible in comparison with the pressure gradient. This means that for this motion the soil could be considered to be an ideal continuum. Furthermore, the soil consisted of very rigid quartz grains packed into a very dense array over geologic time. Therefore, in a well-justified approximation it can be assumed that the soil cannot be consolidated further by repacking. The compressibility of quartz grains is only of significance at pressures of the order of hundreds of thousands of bars. These pressures are at least one order of magnitude higher than the pressures ordinarily obtained in explosions. Therefore the ideal incompressible fluid approximation will give appropriate intermediate asymptotics when one is considering the problem of an explosion in soil.

In a certain approximation, of course, the explosion will create pressure pulses according to the law (4.1). Specialists are assumed to know how much explosive is needed to achieve the prescribed pulse intensity, and how to place it.

If all the points of the body under consideration (the soil mass) have identical velocities, given by relation (4.5), after the explosion then the velocity potential ϕ, with $\partial_1\phi = U_1$, $\partial_2\phi = U_2$, $\partial_3\phi = U_3$, can be obtained by integration and is found to be the linear function

$$\phi(\mathbf{x}, t_0 + 0) = U\cos\alpha\cos\beta x_1 + U\cos\alpha\sin\beta x_2 + U\sin\alpha x_3 + \text{const} . \qquad (4.6)$$

Therefore the pressure pulse intensity, which determines the amount of explosive needed at every point of the mass boundary, must be equal to

$$\theta(\mathbf{x}) = -\rho(U\cos\alpha\cos\beta x_1 + U\cos\alpha\sin\beta x_2 + U\sin\alpha x_3 + \text{const}) . \qquad (4.7)$$

This simple expression seems puzzling because there is an unphysical dependence of the pulse intensity on the choice of coordinate origin; also, the constant is unknown. Both these unphysical properties can be removed by the following argument. It is clear that explosion specialists in explosion techniques can create positive, in fact non-negative, pressure pulses. Therefore we must require that the intensity of the pressure pulse given by relation (4.7) be non-negative at all points of the soil mass boundary. Furthermore, we require that on at least one point of the boundary the pulse intensity be equal to zero: there is no need to spend expensive explosive pointlessly. These conditions allow us to determine the value of the constant that enters expression (4.7) and remove a seeming dependence of the pulse intensity on the choice of origin of the coordinate system. This concludes the solution to the problem.

This directed explosion is a remarkable example of a situation where the correct formulation of the problem was crucial; when this was achieved the answer was obtained easily by elementary arguments rather than the application of a sophisticated technique.

4.2 Lift force on a wing

4.2.1 Physical formulation of the problem. The basic model of the phenomenon. Possibility of using the ideal incompressible fluid approximation

Aviation, doubtless one of the greatest achievements in engineering, entered the scene of technology at the beginning of the last century. Although it was predicted by Leonardo da Vinci, it entered rather unexpectedly, and its potential importance

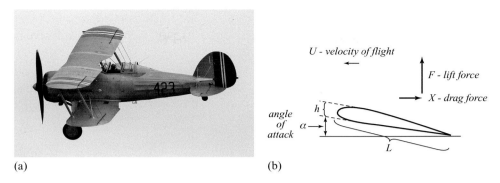

Figure 4.2 (a) An aircraft of the WWII era. (b) The wing of a heavier-than-air flying machine: a principal scheme.

was not properly understood even by outstanding minds, as is illustrated by just two opinions from figures of stature and influence:

Heavier-than-air flying machines are impossible.
(Lord Kelvin, President of the Royal Society of London, 1895)
Airplanes are interesting toys but of no military value.
(Marechal Foch, at that time Professor of Strategy, l'École Supérieure de Guerre, future Commander-in-Chief of the joint Anglo–French forces during World War I)

Already we have learned enough to understand what supports heavier-than-air flying machines – the lift force. The ideal incompressible fluid approximation allows us to calculate the lift force on the wings of World War II aircraft (Figure 4.2a), which were able to reach speeds of 300–400 km/hour. For higher speeds the effects of the compressibility of air are substantial.

Furthermore, aircraft wings are slender: the ratio of their thickness h to their characteristic longitudinal length size L (the "chord" in Figure 4.2b) is a small quantity. Moreover, the "angle of attack" α of the wing to the flight direction is also small. Experiment shows that, for small angles of attack, slender wings and sufficiently large flight speeds, the shear stresses are negligibly small everywhere except at a thin layer near the wing, the so-called "boundary layer", the modeling of which will be considered later.

This may be seen from the photograph in Figure 4.3. The dye-marked fluid moves around the wing. It is clear that in the region immediately adjacent to the wing surface the velocity gradient is large and, because the shear stress is proportional to this gradient, it too is large. Outside this region the velocity gradient is small, and so is the shear stress. It can be shown (see Chapter 8) that the pressure drop across the boundary layer is small, so that pressure distribution across the boundary layer can be assumed constant. The lift force is a result of the air pressure

Figure 4.3 Photograph of the flow around a wing; the visualization was achieved by marked particles. From Prandtl and Tietjens (1931) with kind permission of Springer Science+Business Media, Abbildung 47.

difference between the bottom and top of the wing. We repeat: when calculating the lift force it is possible to neglect, in the first approximation, the existence of the boundary layer and to apply the approximation of an *ideal* fluid.

We emphasize that the ideal fluid approximation is inappropriate for determining the wing drag, i.e. the force component acting on the wing in the direction of the velocity. This component is determined to a large extent by the tangential stresses at the wing surface. The ideal fluid approximation is also inappropriate for determining the lift force at a large angle of attack or when the ratio h/L is not small. In these cases a vortex zone is formed at the rear of the wing, and we would need a better approximation than that of an ideal fluid to calculate the pressure distribution in this zone.

Now, remembering that the wing chord is L and that the velocity is U, we see that the term $d\rho/dt$ in the continuity equation is of order $U\Delta\rho/L$. Here $\Delta\rho$ is a characteristic density variation, equal to $(\Delta\rho/\Delta p)\Delta p$. The motion of the wing is sufficiently fast for us to neglect the heat exchange from the air particles as they move along the wing. Therefore $\Delta\rho/\Delta p \sim 1/a^2$, where $a = \sqrt{dp/d\rho}$ is the velocity of sound in air at rest, of order 1000 km/hr. The largest terms of the sum $\rho\partial_\alpha u_\alpha$ in the continuity equation are of order $\rho\, U/L$. Therefore the ratio of the term $d\rho/dt$ and the largest terms of the sum $\rho\,\text{div}\,\mathbf{u}$ in the continuity equation is of order $U^2/a^2 = M^2$, where M is the Mach number (see Chapter 2), the ratio of the wing's velocity and the sound velocity. For old aircraft $M^2 \sim 0.1$, so the term $d\rho/dt$ is small in comparison with the largest terms of the sum $\rho\,\text{div}\,\mathbf{u}$ and can be neglected. Therefore to understand the principles of the phenomenon the incompressible fluid approximation can be used.

To undertand these principles we will simplify further the shape of the wing and, ignoring the tip effects, consider a "wing of infinite span", i.e. a cylinder having wing cross-section profile C and directrices perpendicular to the direction of flight.

Therefore we can consider the motion as planar. Undoubtedly, to a certain degree this assumption is valid for the middle part of a real wing (see Figure 4.2a).

The motion of the wing is considered to be steady and translational. We assume also that flight is at a high altitude so that the flow region can be considered to be infinite. The density of air is assumed to be constant, so in addition gravitational effects can be neglected.

We emphasize, and it will become very important later, that the steady state of the motion is not reached immediately, but after some transitional period, and also that the viscosity of air, although small, is nevertheless finite.

4.2.2 Formulation of the mathematical problem

We now attach a reference system to the wing; in this frame the flow of air around the wing can be considered steady. We will bear in mind that the field of steady motion in which we are interested is an asymptotic state reached at sufficiently large times. In the early, unsteady, stage of a flight, soon after take-off, this mathematical model is inappropriate. The motion originates from a state of rest; therefore in the approximation of an ideal incompressible fluid it is a potential one.

Thus, in calculating the lift force we have to determine the planar steady potential flow around a cylindrical wing having *profile C*, in the ideal incompressible fluid approximation. The analysis is performed in a Cartesian orthonormal system $x_1 x_2$ attached to the profile. To determine the appropriate complex potential $w(z)$, an analytic function of the complex variable $z = x_1 + ix_2$, the following boundary conditions are formulated naturally.

The first condition is the condition at infinity, i.e. far from the wing where it does not disturb the air motion. Therefore, as x_1 and x_2 go to infinity, $u_1 \to U_1$ and $u_2 \to U_2$, where U_1 and U_2 are components of the velocity of the wing with opposite signs. So, the condition at infinity is obtained in the form

$$(dw/dz)_{z=\infty} = U_1 - iU_2 \qquad (4.8)$$

(we remind the reader that $u_1 + iu_2 = \overline{dw/dz}$).

The second boundary condition takes into account that the air velocity is directed along the wing surface. We reiterate that the boundary layer where the approximation of an ideal fluid is not valid is thin and so the pressure drop across this layer is negligible; therefore we can neglect the existence of this layer in determining the lift force. This means that the boundary condition is imposed not at the outer edge of the boundary layer but at the wing surface itself. Thus the contour ∂C of the wing profile should be part of a stream line. This determines the second condition for the complex potential on the contour ∂C:

$$\psi = \operatorname{Im} w(z) = \text{const} . \qquad (4.9)$$

4.2.3 Construction of a solution

It is convenient to move from the physical complex plane z to a "parametric" complex plane ζ in the following way. Introduce an analytic function $\zeta = F(z)$ that maps conformally the exterior of the wing profile C onto the exterior of a circle γ of radius R. If we require that, in this mapping, points at infinity go to points at infinity and also that the condition

$$\left(\frac{dz}{d\zeta}\right)_{\zeta=\infty} = 1 \tag{4.10}$$

be valid then the mapping is fully determined (as is known from the theory of functions of a complex variable) and, in particular, the radius R is determined uniquely. Denote by $f(\zeta) : z = f(\zeta)$ the function inverse to $F(z)$. We see that both f and F are single-valued analytic functions having a simple pole at infinity. Introduce the function $W(\zeta) = w[f(\zeta)]$. It should be clear that the function $W(\zeta)$ is an analytic function of a complex variable corresponding to a certain flow of an ideal incompressible fluid in the parametric plane ζ. We have

$$\frac{dW}{d\zeta} = \frac{dw}{dz}\frac{dz}{d\zeta} \ .$$

Therefore the boundary conditions for the corresponding flow in the parametric plane ζ assume the form

$$\left(\frac{dW}{d\zeta}\right)_{\zeta=\infty} = U_1 - iU_2 \qquad (\operatorname{Im} W)_{\partial\gamma} = \text{const} \ ; \tag{4.11}$$

here $\partial\gamma$ is the circular boundary of the circle γ.

The function $W(\zeta)$, which is analytic outside γ, is determined up to a real constant by the conditions formulated above. Indeed, represent the function W as the sum

$$W = W_1 + W_2 + W_3 \ ,$$

so that

$$\left(\frac{dW_1}{d\zeta}\right)_{\zeta=\infty} = U_1 \ , \qquad (\operatorname{Im} W_1)_{\partial\gamma} = \text{const} \ ,$$

$$\left(\frac{dW_2}{d\zeta}\right)_{\zeta=\infty} = -iU_2 \ , \qquad (\operatorname{Im} W_2)_{\partial\gamma} = \text{const} \ , \tag{4.12}$$

$$\left(\frac{dW_3}{d\zeta}\right)_{\zeta=\infty} = 0 \ , \qquad (\operatorname{Im} W_3)_{\partial\gamma} = \text{const} \ .$$

To determine the function W_1, note that the conformal mapping $\zeta_1 = \zeta + R^2/\zeta$ transforms the exterior of the circle γ of radius R onto the exterior of the cut

$-2R \leq x_1 \leq 2R$ of the real axis. Evidently, $\left(\frac{d\zeta_1}{d\zeta}\right)_\infty = 1$, $\left(\frac{dW_1}{d\zeta}\right)_\infty = \left(\frac{dW_1}{d\zeta_1}\right)_\infty = U_1$. The problem is reduced to determining the complex potential of a flow around a straight line parallel to the velocity of the flow.

We note now that in the ideal incompressible fluid approximation it is possible to replace any stream line by a material line; the flow field will remain unchanged. In particular, if in a uniform translational flow having constant velocity everywhere we replace part of any rectilinear stream line by a solid straight material line then the translational flow will remain as it was. Therefore the solution to the problem of uniform flow around a solid straight line in the plane ζ_1 with velocity U_1 is obtained trivially:

$$W_1 = U_1 \zeta_1 = U_1 \left(\zeta + \frac{R^2}{\zeta}\right). \tag{4.13}$$

Similarly the mapping $\zeta_2 = \zeta - R^2/\zeta$ transforms the exterior of the circle γ onto the exterior of the cut $-2R \leq x_2 \leq 2R$ along an imaginary axis. Therefore determining W_2 is reduced to the solution of a similar trivial problem of uniform flow around a straight line parallel to the flow velocity. We obtain

$$W_2 = -iU_2 \zeta_2 = -iU_2 \left(\zeta - \frac{R^2}{\zeta}\right). \tag{4.14}$$

Furthermore, as can be shown by considering the solution for a vortex obtained earlier, the potential

$$W_3 = \frac{\Gamma}{2\pi i} \ln \zeta, \tag{4.15}$$

where Γ is an arbitrary real number, also corresponds to all the imposed conditions: we have $(dW_3/d\zeta)_{z=\infty} = 0$ and $(\mathrm{Im}\, W_3)_{\partial\gamma} = \mathrm{const}$. Thus, the function

$$W = (U_1 - iU_2)\zeta + (U_1 + iU_2)\frac{R^2}{\zeta} + \frac{\Gamma}{2\pi i} \ln \zeta \tag{4.16}$$

satisfies all the conditions of the problem under consideration. Moreover, this is the most general solution because, as it is easy to show, the function W_3 corresponding to $\Gamma = 0$ is identically equal to zero.

Returning to the complex variable z, we obtain the final result:

$$w(z) = (U_1 - iU_2)F(z) + (U_1 + iU_2)\frac{R^2}{F(z)} + \frac{\Gamma}{2\pi i} \ln F(z). \tag{4.17}$$

Nowadays, in the computer era, determining a function $F(z)$ that conformally maps the exterior of a profile ∂C onto the exterior of a circle γ is a routine matter. Therefore to obtain the complete solution to the problem we need only find one

Figure 4.4 Photograph of the flow around a circular cylinder: a large asymmetric vortex zone behind the cylinder can be seen. From Prandtl and Tietjens (1931) with kind permission of Springer Science+Business Media, Abbildung 6.

parameter, the real number Γ, i.e. the circulation of the vortex which entered our solution rather unexpectedly but has in fact a deep physical meaning.

If we formally put $F(z) = z$, we will obtain the solution for the "circulation flow" around a cylinder in the ideal incompressible fluid approximation:

$$w(z) = (U_1 - iU_2)z + (U_1 + iU_2)\frac{R^2}{z} + \frac{\Gamma}{2\pi i}\ln z. \tag{4.18}$$

This solution does not correspond to a real flow around a cylinder, as can be seen from Figure 4.4. Thus the ideal incompressible fluid approximation, which we used here, is not valid for the flow as a whole owing to the formation of a vortex wake behind the cylinder. Nevertheless this solution is very important in constructing the mathematical model with which we are dealing, and we will use it in subsequent investigations.

4.2.4 Selection of a unique solution

We have obtained a non-unique solution of the problem that we formulated in Section 4.2.2. There is in fact a family of solutions, parametrized by the circulation Γ of the vortex. We need to select a single solution from this family and, moreover, to explain the physical meaning of this vortex and its circulation. The non-uniqueness of the solutions is a common property of nonlinear problems in the mechanics of continua. To make the solution unique we need to use arguments additional to those included in the formulation given in Section 4.2.2. These arguments are based on the analysis of problems whose asymptotics should be the solution we seek. Such an approach to nonlinear problems in the mechanics of continua is very often the only way to force uniqueness.

In the case under consideration the solution we are searching for is in fact doubly asymptotic. First, this family of solutions is obtained by using the ideal fluid approximation whereas in fact the viscosity of air is small but finite. Furthermore, the required solution is in fact an intermediate asymptotic representation of unsteady solutions describing the transition from the state of rest to the state of steady flow.

The analysis of unsteady flows shows that in the transitional regimes a "vortex sheet" appears (Figure 4.5) behind the trailing edge of the wing. In this vortex sheet we can see a sharp variation of the fluid velocity in the transverse direction. At the end of the sheet a vortex forms, which moves into the fluid body. It is here that viscous forces come into play. Owing to viscous dissipation the sharp transverse velocity variation across the vortex sheet decays, as does the vortex at the end of this sheet. The limiting solution is characterized (Figure 4.3) by the smooth structure of the velocity field around the trailing edge, i.e. the smooth departure of air particles from the trailing edge of the wing. Therefore the velocity at the trailing edge must be continuous and finite. It so happens that in the family of solutions given by relation (4.18) there exists only one solution satisfying this condition. Indeed, we have

$$\frac{dw}{dz} = \frac{dW}{d\zeta} : \frac{dz}{d\zeta} .$$
(4.19)

At the point $\zeta = \zeta_0 = Re^{i\theta_0}$, which corresponds to the trailing edge of the wing, and only at this point, the conformity of the mapping $z = f(\zeta)$ is evidently violated: the angle π is transformed in this mapping to an angle larger than π because the trailing edge is sharpened. Therefore at this point $dz/d\zeta = 0$. Thus the derivative dw/dz of the complex potential, and consequently the flow velocity, can be a finite quantity only if $dW/d\zeta$ also vanishes at the point $\zeta = \zeta_0$. Therefore the following condition must be satisfied:

$$\left(\frac{dW}{d\zeta}\right)_{\zeta=\zeta_0} = U_1 - iU_2 - \frac{(U_1 + iU_2)R^2}{\zeta_0^2} + \frac{\Gamma}{2\pi i\zeta_0} = 0 .$$
(4.20)

Without loss of generality we can use a reference system where $U_1 = U$ and $U_2 = 0$ (it can always be achieved by a rotation of the reference system). In this case we obtain from (4.20)

$$\Gamma = -4\pi RU \sin\theta_0 .$$
(4.21)

Within the strict context of the ideal incompressible fluid approximation this condition should be considered as an additional postulate. In the literature it is referred to as the Kutta (sometimes Kutta–Joukovsky) condition or postulate. We emphasize that establishing this condition required an extension of the original model based

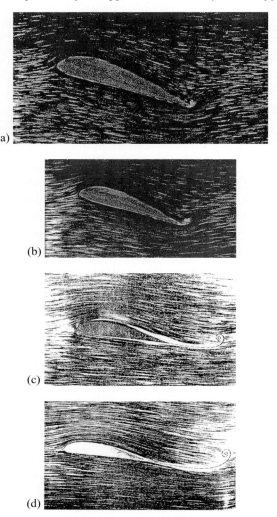

Figure 4.5 The vortex sheet behind the trailing edge of a wing, terminated by a vortex, appears and decays in the transitional regime. (a), (b) The vortex is formed at the trailing edge. (c), (d) The propagation and decay of the vortex and the vortex sheet attached to the vortex. From Prandtl and Tietjens (1931) with kind permission of Springer Science+Business Media, Abbildungen 43–6.

on assuming a steady ideal flow. The name of the postulate requires a historical discourse. In fact, this postulate was formulated and substantiated simultaneously and independently by the German applied mathematician W. M. Kutta (1910) and the Russian applied mathematician S. A. Chaplyguine (1910). It is very important to note that in both these articles the necessity for a rounded leading edge and a sharp trailing edge of the profile was emphasized. Chaplyguine's paper was published in Russian in a Russian mathematical journal, whereas Kutta's was published in German. Soon after these papers there appeared in German, in a leading aeronautical journal of that time, a paper by N. E. Joukovsky (1910) where the results of both Chaplyguine and Kutta were presented, so that the work of each of these scientists was explained and emphasized. Also, Joukovsky compared their results with an earlier paper of Kutta (1902) in which the special case of a circulation flow around a symmetric profile having the shape of a circular arc was considered. In that paper Kutta demonstrated that, for a certain value of the circulation, the velocity at either end of the arc is finite, and he also calculated the lift force for this case. (This solution is structurally unstable because any, even the smallest, violation of symmetry leads to the formation of a concentrated force at the leading or trailing edge, and so it is not related to the theory of flight.) In his 1910 paper Joukovsky also presented the mathematical construction of a family of wing profiles (now called Joukovsky profiles) that later became popular in aircraft design. So, it would be more appropriate to name the condition (4.21) the Kutta–Chaplyguine–Joukovsky postulate.

4.2.5 Calculation of the lift force

We turn now to a calculation of the lift force acting on a wing. We first calculate the forces which would act on a circular cylinder in a uniform flow with velocity U at infinity. In an imaginary case we can use the ideal incompressible fluid approximation for the flow around the cylinder. From (4.19) the complex potential for such a flow would be ($U_1 = U$, $U_2 = 0$)

$$w(z) = U\left(z + \frac{R^2}{z}\right) + \frac{\Gamma}{2\pi i}\ln z. \qquad (4.22)$$

It is clearly seen (Figure 4.6) that the components of the force acting on a unit width of the cylinder have the following expressions:

The force component F, perpendicular to the velocity, i.e. the *lift force*, is given by

$$F = -\int_{\partial\gamma} p\sin\theta\, ds = -\int_0^{2\pi}(p\sin\theta)\, R\, d\theta; \qquad (4.23)$$

the force component X, parallel to the velocity, i.e. the *drag force*, is given by

$$X = \int_{\partial\gamma} p\cos\theta\, ds = -\int_0^{2\pi}(p\cos\theta)\, R\, d\theta. \qquad (4.24)$$

The integrals are taken over the cylinder surface, $\partial\gamma$, so that $ds = R\, d\theta$. The pressure can be expressed via the Lagrange–Cauchy integral (3.28), which coincides in

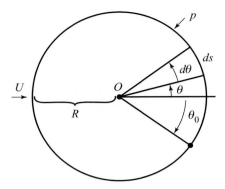

Figure 4.6 Calculation of the lift force and the drag force in a fictitious ideal incompressible flow around a circular cylinder.

this case with the Bernoulli integral: thus

$$p = C - \frac{\rho \mathbf{u}^2}{2} ,$$

where C is a constant and

$$\mathbf{u}^2 = u_1^2 + u_2^2 = (u_1 + iu_2)(u_1 - iu_2) = \left(\frac{dw}{dz}\right)\overline{\left(\frac{dw}{dz}\right)} .$$

We find from (4.23) and these formulae that

$$\mathbf{u}^2 = 4U^2 \sin^2 \theta - \frac{2\Gamma U}{\pi R} \sin \theta + \frac{\Gamma^2}{4\pi^2 R^2} . \tag{4.25}$$

Substituting (4.25) into the expression for the pressure and then performing the integrations in (4.23), (4.24) we obtain for the cylinder

$$F = -\rho U \Gamma , \qquad X = 0 . \tag{4.26}$$

Now we will determine the force acting on a wing of an arbitrary profile, assuming the ideal incompressible fluid approximation. We remind the reader that for an ideal incompressible fluid the components of the tensor of momentum flux density $\mathbf{\Pi} = p\mathbf{I} + \rho \mathbf{u} \otimes \mathbf{u}$ have the following expression: $\Pi_{ij} = p\delta_{ij} + \rho u_i u_j$. The flow is steady and the mass force vanishes. Therefore the equation for momentum balance takes the form

$$\operatorname{div} \mathbf{\Pi} = 0 , \qquad \partial_\alpha \Pi_{i\alpha} = 0.$$

Take a contour ∂S which is far from the wing contour (Figure 4.7). This contour will be removed to infinity later. Integrating the relation $\partial_\alpha \Pi_{i\alpha} = 0$ over the volume

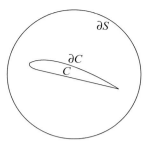

Figure 4.7 Calculation of the lift force due to the flow around a wing.

between contours ∂C and ∂S fixed in the reference system attached to the wing and using Gauss' theorem, we obtain

$$\int \partial_\alpha \Pi_{i\alpha}\, d\omega = \int_{\partial C} \Pi_{i\alpha} n_\alpha\, ds + \int_{\partial S} \Pi_{i\alpha} n_\alpha\, ds \,. \qquad (4.27)$$

Here n_α is a component of the outward normal to the boundary of the volume of integration. We have

$$\int_{\partial C} \Pi_{i\alpha} n_\alpha\, ds = \int_{\partial C} (\rho u_i u_\alpha + p\delta_{i\alpha}) n_\alpha\, ds = \int_{\partial C} p n_i\, ds \,, \qquad (4.28)$$

because $n_\alpha \delta_{i\alpha} = n_i$ and there is no flux through the contour ∂C, so that $\rho u_\alpha n_\alpha = 0$ at ∂C. It is easy to see that the last integral in (4.28), taken with a minus sign, is the ith component I_i of the force acting on unit width of the wing. Thus

$$I_i = -\int_{\partial C} p n_i\, ds = \int_{\partial S} (p\delta_{i\alpha} + \rho u_i u_\alpha) n_\alpha\, ds \,. \qquad (4.29)$$

Note that as $z \to \infty$, $d\zeta/dz \to 1$ and the conformal mapping $\zeta = F(z)$ can be represented by

$$\zeta = F(z) = z\left(1 + O\left(\frac{1}{z}\right)\right)$$

so that, as $z \to \infty$,

$$w = Uz + \frac{\Gamma}{2\pi i} \ln z + O\left(\frac{1}{z}\right), \qquad \frac{dw}{dz} = U + \frac{\Gamma}{2\pi i z} + O\left(\frac{1}{z^2}\right). \qquad (4.30)$$

We note that the terms of order $O(1/z^2)$ are negligible in calculating the integral (4.29) when the contour ∂S goes to infinity. Therefore we come to an important conclusion: *the specific shape of the wings influences the force acting on the wings only via the circulation* Γ. Thus both components of the total force **I**, i.e. the lift force $I_2 = F$ and the force $I_1 = X$, acting on the profile C in the ideal incompressible fluid approximation depend only on the parameters U, Γ and ρ. The dimensions of F and X are M/T^2 (force per unit length); the dimensions of U, Γ and ρ are L/T, L^2T and M/L^3. Dimensional analysis therefore gives

$$F = c_1\rho U\Gamma\,, \qquad X = c_2\rho U\Gamma, \tag{4.31}$$

where c_1, c_2 are constants that are universal in the sense they do not depend on the shape of the wing profile; remember that the shape of the wing influences the force only via the circulation Γ. The direct computation performed above for the case of a circular cylinder yielded $c_1 = -1$, $c_2 = 0$, and these values should be valid for a wing of arbitrary profile. Therefore we obtain, in the ideal incompressible fluid approximation, the same expression as (4.26) for a wing of arbitrary profile (4.27):

$$F = -\rho U\Gamma\,, \qquad X = 0\,. \tag{4.32}$$

In the literature the first of these relations is called the *Joukovsky theorem* and the second the *D'Alembert paradox*. The latter term arose for historical reasons and is not quite correct: the vanishing of the drag in the ideal incompressible fluid approximation cannot be considered a paradox because this approximation, as shown earlier, is inappropriate for calculating the drag force.

The Joukovsky (1906) theorem[3] shows that the lift force is proportional to the product of the velocity of flight U and the circulation Γ. We calculated the circulation Γ in the ideal incompressible fluid approximation using the Kutta–Chaplyguine–Joukovsky condition. However, the origin of the circulation is related to the production of vorticity at the boundary by the action of viscous forces, which are not taken into account in this approximation.

Note in conclusion that for a symmetric profile under zero angle of attack the circulation, by symmetry, is equal to zero. At small angles of attack the circulation is proportional to the angle of attack. Therefore the lift force is then proportional to the product of the velocity and the angle of attack: $F \sim \rho U\alpha$.

---

[3] It is fair to mention that the Joukovsky theorem for the case of a circular cylinder was substantiated, formulated and proved in Lord Rayleigh's (1877) paper "On the irregular flight of a tennis ball".

5

The linear elastic solid approximation. Basic equations and boundary value problems in the linear theory of elasticity

The mathematical theory of elasticity based on the idealization presented in this chapter is a remarkable classical branch of the mechanics of continua, which has advanced far in the more than 200 years it has been studied. In this chapter a concise presentation of the fundamentals of this theory will be given, bearing in mind readers who have not met it before, and it will also serve as a preparation for the next chapter, where we discuss the mathematical modeling of fracture phenomena, which nowadays is the principal area of attention.

5.1 The fundamental idealization

A crucially important property of a deformable solid continuum is that it is possible for it to possess non-trivial stress distributions even when the body is at rest, i.e. when the velocity is everywhere equal to zero.

The theory of elasticity as a science is older than fluid mechanics. Its basic law, which was developed to a fundamental model, was formulated by Robert Hooke more than 300 years ago, in the article Hooke (1678).[1]

Readers already know the formulation of Hooke's law for an elastic rod. A rod[2] is an elastic body whose length ℓ is substantially larger than its cross-sectional size s and which has a constant cross-section area S (see Figure 5.1, taken from the book of Galileo Galilei (1638)). Let us take the longitudinal direction of the rod as the x_1 axis of a system of orthonormal Cartesian coordinates. Then, under the

---

[1] Before the paper of 1678, Hooke announced in 1660 at the end of his *Book of the Descriptions of Helioscopes* this law in the form of an anagram in Latin: *c e i i i n o s s s t t u v – ut tensio, sic vis*, which means "as I stretch, so is the force". At that time such a form of publishing was not unusual and, for mathematicians of the time, solving such anagrams was not a big deal. However, the paper by Hooke (1678) should be considered as the first publication in the theory of elasticity. Hooke wrote afterwards that he omitted to publish the law earlier because he was anxious to obtain a patent for a particular application of it (Todhunter and Pearson, 1886).

[2] In fact, this concept of a rod is also an example of an "intermediate asymptotic". Speaking about deformations of a rod we have in mind deformations in a region (Figure 5.1) between the grips (supports) at distances from the grips much larger than the cross-sectional size.

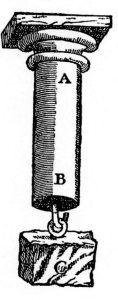

Figure 5.1 Galilei's illustration of a tensile test.

weight P of the load C hanging on its grip and directed along its axis AB, the rod is *strained*: it is stretched and thinned. The amounts of stretching $\Delta\ell$ and thinning Δs are related to the acting loading force P by Hooke's law:

$$\frac{\Delta\ell}{\ell} = \frac{P}{SE}, \qquad \frac{\Delta s}{s} = -\frac{\nu P}{SE}. \tag{5.1}$$

Here E and ν are the material properties, the *Young modulus* and the *Poisson ratio* respectively. The real content of the law is that for an elastic body these properties depend neither on the size of the rod nor on the load, so that if for a given material these properties are measured, it is possible to find deformations for all rods and all loads.

The fundamental idealization, which we will speak about in this chapter, is based on two important observations. First, for a wide class of materials (metals, ceramics, composites etc.) under normal conditions and moderate loads the value of the Young modulus is very large, much larger than the stress P/S which the material can support. For instance, for structural steel E is of the order of 10^6 kgf/cm^2 whereas normally loads are of the order of 10^3 kgf/cm^2. Therefore the relative stretching and thinning $\Delta\ell/\ell$, $\Delta s/s$ are small; e.g. for a steel rod 1 m long and 1 cm diameter, loaded by 1 ton, a substantial load, the stretching is of the order of a millimeter and the thinning is of the order of 10 microns, i.e. the thickness of a human

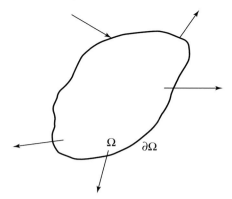

Figure 5.2 Elastic body under "generic" loads.

hair. Furthermore, when a load, even a substantial one, is removed from structures made of such materials, they return to their previous undeformed state.

Therefore, under a very wide range of external conditions (temperature, ambient medium etc.) and loads, structures made of such materials can be approximately modeled as *linear elastic bodies*. The relations between forces and deformations in this model essentially extend those of Hooke's law (5.1) to the general case of spatial loadings and geometries. It is significant that in such generalizations we need introduce no new material properties other than the Young modulus E and the Poisson ratio v.

To obtain the basic relationships for the linear elastic body approximation con-sider the deformation of an arbitrary body under the action of some "generic" loads (Figure 5.2). The first assumption is that the continuous medium is homogeneous and isotropic: the properties of the body do not depend on the origin of the refer-ence system of the observer nor on its orientation.

We will proceed using the Lagrangian approach, introducing the time t and initial radius vector of a particle, $\mathbf{X}$, as independent variables. After the application of loads the particles forming the deformable body are displaced, so that the particle with radius vector $\mathbf{X}$ at time t_0 just before the loading is, at time t, moved to the radius vector $\mathbf{x}$, $\mathbf{x} = \chi(\mathbf{X}, t)$. In the approximation under consideration it is assumed that the vector $\mathbf{w}$ of the deformation displacement,

$$\mathbf{w}(\mathbf{X}, t) = \mathbf{x} - \mathbf{X},\qquad(5.2)$$

has a small magnitude (modulus) in comparison with the characteristic length scale ℓ of the body. More precisely, introduce the characteristic stress $\sigma_0 = P/\ell^2$, where P is a characteristic load acting on the body. Then

$$\varepsilon = \frac{\sigma_0}{E}$$

is a small parameter. We assume that the magnitude of the displacement vector $|\mathbf{w}|$ is of the order of $\varepsilon\ell$ over the entire body. Note that this is a special case, that of stiff elastic bodies. Large displacements can be considered for perfectly elastic bodies (rubber is an example), but this is beyond the scope of the present book.

Therefore, in the present model we are justified in neglecting the difference between the Eulerian and Lagrangian independent variables, $\mathbf{x}$ and $\mathbf{X}$, in the equation of momentum balance (1.26), and so we identify in this equation the Eulerian independent variable $\mathbf{x}$, used in its derivation, with the Lagrangian independent variable $\mathbf{X}$, the radius vector of a particle prior to loading.

Now take a small neighborhood of the point $\mathbf{x}$ and expand the field of the deformation displacement $\mathbf{w}(\mathbf{x} + \mathbf{r})$ in this vicinity in a Taylor series, leaving only the first two terms (we will see later what this means physically):

$$\mathbf{w}(\mathbf{x} + \mathbf{r}) = \mathbf{w}(\mathbf{x}) + (\text{grad } \mathbf{w}) \cdot \mathbf{r} . \tag{5.3}$$

Here, grad $\mathbf{w}$ is the second-rank tensor with components $\partial_j w_i$. Again, as we did in Chapter 3, but this time for the displacement, we represent the second-rank tensor grad $\mathbf{w}$ as a sum of symmetric and antisymmetric tensors:

$$\mathbf{L} = \text{grad } \mathbf{w} = \mathbf{W} + \mathbf{D} ,$$
$$\mathbf{W} = \tfrac{1}{2}(\text{grad } \mathbf{w} - \text{grad } \mathbf{w}^{\mathrm{T}}) , \tag{5.4}$$
$$\mathbf{D} = \tfrac{1}{2}(\text{grad } \mathbf{w} + \text{grad } \mathbf{w}^{\mathrm{T}}) .$$

Thus the vector $(\text{grad } \mathbf{w}) \cdot \mathbf{r}$ is equal to the sum of the vectors $\mathbf{W} \cdot \mathbf{r} + \mathbf{D} \cdot \mathbf{r}$. The former, $\mathbf{W} \cdot \mathbf{r}$, is equal to the vector product $\boldsymbol{\omega} \times \mathbf{r}$, where (see Chapter 3) $\boldsymbol{\omega} = \tfrac{1}{2} \text{ curl } \mathbf{w}$ is the axial vector corresponding to the antisymmetric tensor W and is called the *small-rotation vector*. Therefore the sum $\mathbf{w}(\mathbf{x}) + \mathbf{W} \cdot \mathbf{r}$ represents the displacement of particles in the neighborhood of a point $\mathbf{x}$ in a solid body: it is a translational motion with center $\mathbf{w}(\mathbf{x})$ and a small rotation $\boldsymbol{\omega} \times \mathbf{r}$. In orthogonal Cartesian coordinates the tensor $\mathbf{D}$ has components

$$D_{ij} = \tfrac{1}{2}(\partial_j w_i + \partial_i w_j) . \tag{5.5}$$

It is easy to show that if the displacement field $\mathbf{w}(\mathbf{x} + \mathbf{r})$ corresponds to the displacement of a solid body, so that

$$\mathbf{w}(\mathbf{x} + \mathbf{r}) = \mathbf{w}(\mathbf{x}) + \boldsymbol{\Omega} \times \mathbf{r}$$

where $\boldsymbol{\Omega}$ is a constant vector, then the tensor $\mathbf{D}$ with components $\partial_j w_i + \partial_i w_j$ vanishes. Therefore this tensor expresses the contribution of what remains – of the local small deformation – to the displacement field. This explains its name: the *small-strain tensor.*

The tensor $\mathbf{D}$ is symmetric and so has three mutually orthogonal principal directions, determined by the unit vectors $\mathbf{e}_1$, $\mathbf{e}_2$, $\mathbf{e}_3$. We can, without any loss of

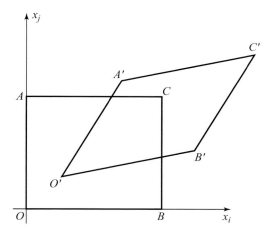

Figure 5.3 Deformation of an infinitesimal rectangle $OACB$ to the parallelogram $O'A'C'B'$: $O, (x_i, x_j); A, (x_i, x_j+dx_j); B, (x_i+dx_i, x_j); C, (x_i+dx_i, x_j+dx_j); O', (x_i+w_i, x_j+w_j); A', (x_i+w_i+\partial_j w_i dx_j, x_j+dx_j+w_j+\partial_j w_j dx_j); B', (x_i+dx_i+w_i+\partial_i w_i dx_i, x_j+w_j+\partial_i w_j dx_i); C', (x_i+dx_i+w_i+\partial_i w_i dx_i+\partial_j w_i dx_j, x_j+dx_j+w_j+\partial_i w_j dx_i+\partial_j w_j dx_j)$.

generality, select the local system x_1, x_2, x_3 in such a way that the three vectors $\mathbf{e}_1$, $\mathbf{e}_2$, $\mathbf{e}_3$ form its basis. Therefore the action of a small deformation is reduced to stretching or contracting the small vectors $dx_i \mathbf{e}_i$ to $dx'_i \mathbf{e}_i$ ($i = 1, 2, 3$, note that there is no summation!) with relative strain $(\Delta \ell / \ell) = (dx'_i - dx_i)/dx_i = \partial_i w_i$, while these vectors remain orthogonal. Thus, a rectilinear parallelepiped based on these vectors remains a rectilinear parallelepiped. The relative variation of the volume of this parallelepiped, $\Delta V / V$, where $V = dx_1 dx_2 dx_3$, $\Delta V = V' - V$ and $V' = dx'_1 dx'_2 dx'_3$, is equal, up to terms of the next order, to the trace of the tensor $\mathbf{D}$:

$$\frac{\Delta V}{V} = \text{trace } \mathbf{D} = \partial_\alpha w_\alpha \tag{5.6}$$

(here the summation law for indices applies, of course). The result obtained is natural: the quantity trace $\mathbf{D}$ is an *invariant* and so remains unchanged by any rotation of the local coordinate system.

The physical meaning of the components D_{ij} of the tensor $\mathbf{D}$ in an arbitrary system of coordinates is clear from Figure 5.3. For $i = j$, up to terms of the next order it is the relative stretching (or contraction) of a small vector dx_i directed along the axis x_i; for $i \neq j$ it is one-half the variation, after a small deformation, of what was previously a right angle between the vectors $d\mathbf{x}_i$ and $d\mathbf{x}_j$. The reader is recommended to check this statement; it is a good exercise.

Now introduce the linear elastic deformable solid approximation, for which the stress tensor $\boldsymbol{\sigma}$ is a linear function of the small-strain tensor $\mathbf{D}$. We remind the reader additionally of the assumption that the elastic solid is *homogeneous* and *isotropic.* Then the most general form of linear relation between the tensors $\boldsymbol{\sigma}$ and $\mathbf{D}$ is

$$\boldsymbol{\sigma} = \lambda(\text{trace}\,\mathbf{D})\mathbf{I} + 2\mu\mathbf{D} . \tag{5.7}$$

Here λ and μ are constant scalar properties of the medium, known as the *Lamé constants*. They are constant because if they were to depend on the coordinates then relation (5.7) would be inhomogeneous. They are scalars because if they were tensors they would have some preferred directions: this is incompatible with the isotropy assumption.

The reader may ask why are there no linear terms with higher, e.g. second, derivatives in relations of the type (5.7). The physical reason for this is that were such terms to enter relation (5.7) then the constant coefficients associated with them would have the dimensions of the Lamé constants times one extra length. Dividing these new constants by the Lamé constants we would find new characteristic material linear scales entering the model. Our restriction to first derivatives therefore means that in the intermediate asymptotic model under consideration such length scales are taken to be negligible. Sometimes it is necessary to drop this assumption, for instance in the modeling of short-wavelength elastic waves.

Equation (5.7) is a *constitutive equation* for a *linear elastic homogeneous isotropic body*, and the corresponding approximation is called the *linear elastic body approximation*. It is appropriate to introduce the tensor

$$\mathbf{D} - (\tfrac{1}{3}\,\text{trace}\,\mathbf{D})\mathbf{I} , \tag{5.8}$$

which is called the small-strain *deviator*. The trace of this tensor is obviously equal to zero. In terms of this deviator the constitutive relation (5.7) becomes

$$\boldsymbol{\sigma} = 2\mu[\mathbf{D} - \tfrac{1}{3}(\text{trace}\,\mathbf{D})\mathbf{I}] + K(\text{trace}\,\mathbf{D})\mathbf{I} . \tag{5.9}$$

Here $K = \lambda + \tfrac{2}{3}\mu$. Equation (5.9) is also a widely used form of the relation between the stress tensor and the small-strain tensor – it is the constitutive equation of a homogeneous isotropic linear elastic deformable solid.

Owing to the special importance of relations (5.7)–(5.9) we will give here another derivation of them, based explicitly on the invariance principle.

Let $\mathbf{a}$ and $\mathbf{b}$ be arbitrary vectors. Consider a scalar, the bilinear form $\sigma_{\alpha\beta}a_{\alpha}b_{\beta}$ (the summation is again over repeated Greek indices from 1 to 3). Owing to the linearity, homogeneity and isotropy of the medium, this bilinear form can be interpreted as the most general linear function of three quantities:

(1) the analogous scalar bilinear form for the small-strain tensor $D_{\alpha\beta}a_{\alpha}b_{\beta}$;
(2) a linear invariant of the tensor $\mathbf{D}$, its trace, trace $\mathbf{D} = D_{\alpha\alpha}$; and
(3) the joint bilinear invariant of the vectors $\mathbf{a}$ and $\mathbf{b}$, their scalar product $(\mathbf{a}, \mathbf{b}) = a_{\alpha}b_{\alpha}$.

The most general form of such a linear relation is obviously

$$\sigma_{\alpha\beta}a_\alpha b_\beta = \lambda D_{\alpha\alpha}a_\gamma b_\gamma + 2\mu D_{\alpha\beta}a_\alpha b_\beta$$

(the scalar product $a_\alpha b_\alpha$ cannot enter separately because if $\mathbf{D} = 0$ then $\boldsymbol{\sigma}$ would also be equal to zero; we will not consider a so-called "pre-stressing" situation here).

Assuming $\mathbf{a} = \mathbf{e}_i$, $\mathbf{b} = \mathbf{e}_j$ (where $\mathbf{e}_1, \mathbf{e}_2, \mathbf{e}_3$ are the unit vectors of the Cartesian orthonormal system) we obtain again relation (5.7) in component form,

$$\sigma_{\alpha\beta}a_\alpha b_\beta = \sigma_{ij} = \lambda D_{\alpha\alpha}\delta_{ij} + 2\mu D_{ij} ,$$

and also

$$\sigma_{ij} = 2\mu(D_{ij} - \tfrac{1}{3}D_{\alpha\alpha}\delta_{ij}) + K D_{\alpha\alpha}\delta_{ij} , \tag{5.10}$$

which is relation (5.9) in component form. (The factor 2 compensates for the previously introduced factor $\tfrac{1}{2}$.) The physical meaning of the constants λ and K is transparent. Indeed, the trace of the deviator $\mathbf{D} - \tfrac{1}{3}$ trace $\mathbf{DI}$ vanishes. Therefore the relation (5.9) after convolution (i.e. setting $i = j$ and summing) assumes the form

$$\text{trace } \boldsymbol{\sigma} = 3K \text{ trace } \mathbf{D} . \tag{5.11}$$

Thus the coefficient $3K$ is the proportionality factor between the traces of the stress and small-strain tensors. The latter, trace $\mathbf{D} = \partial_\alpha w_\alpha$, is, as we have just seen, the volume strain. In particular, in the case of a hydrostatic pressure, i.e. if the stress has no shear components, we have $\boldsymbol{\sigma} = -pI$ and trace $\boldsymbol{\sigma} = -3p$ and relation (5.11) assumes the form

$$\text{trace } \mathbf{D} = -\frac{p}{K} . \tag{5.12}$$

Thus the coefficient K is equal to the proportionality factor between the pressure and the volume strain; it is called the *volume compression modulus.*

To understand the meaning of the second coefficient μ let us turn to relation (5.10) and put $i \neq j$. We obtain

$$\sigma_{ij} = 2\mu D_{ij} = \mu(\partial_j w_i + \partial_i w_j) . \tag{5.13}$$

However, the quantity $\partial_j w_i + \partial_i w_j$ (see the caption for Figure 5.3) is just the variation of the angle between two initially mutually perpendicular infinitesimally small vectors $d\mathbf{x}_i$ and $d\mathbf{x}_j$. Thus μ is the proportionality factor between the shear stress acting at the faces normal to $d\mathbf{x}_i$ and $d\mathbf{x}_j$ and the corresponding variation of the right angle between these faces. This justifies the term *shear modulus* for the coefficient μ.

Obviously the elastic constants λ, μ, K could be expressed via the Young modulus and Poisson ratio. To obtain such expressions it is enough to solve the relations (5.10) for D_{ij}. We have

$$D_{ij} = \frac{1}{9K} \sigma_{\alpha\alpha}\delta_{ij} + \frac{1}{2\mu}\left(\sigma_{ij} - \frac{1}{3}\sigma_{\alpha\alpha}\delta_{ij}\right) . \tag{5.14}$$

In the case of the tension in a rod, $\sigma_{11} = P/S$ and all other components of the stress tensor vanish. Substituting into (5.14) we obtain

$$D_{11} = \frac{\Delta\ell}{\ell} = \left(\frac{1}{9K} + \frac{1}{3\mu}\right)\frac{P}{S} = \frac{P}{SE} ,$$
$$D_{22} = \frac{\Delta s}{s} = \left(\frac{1}{9K} - \frac{1}{6\mu}\right)\frac{P}{S} = -\frac{\nu P}{SE} . \tag{5.15}$$

From these relations the relations for λ, μ and K as functions of E and ν follow easily:

$$\mu = \frac{E}{2(1+\nu)} , \qquad K = \frac{E}{3(1-2\nu)} , \qquad \lambda = \frac{\nu E}{(1+\nu)(1-2\nu)} . \tag{5.16}$$

5.2 Basic equations and boundary conditions of the linear theory of elasticity

In the absence of mass forces the equation (1.30) for momentum balance assumes the form

$$\rho\frac{d\mathbf{u}}{dt} = \operatorname{div}\boldsymbol{\sigma} . \tag{5.17}$$

On the left-hand side of equation (5.17) is the individual derivative of the velocity $\mathbf{u}$, i.e. the velocity derivative for a given particle. In Lagrangian variables $\mathbf{X}, t$ this derivative is equal to $\partial_t\mathbf{w}$. However, in the same Lagrangian variables we have $\mathbf{u} = \partial_t\mathbf{x}$, where $\mathbf{x}$ is the radius vector of the particle at time t having Lagrangian coordinate $\mathbf{X}$. The quantity $\mathbf{X}$ is obviously time independent. Therefore we obtain in Lagrangian coordinates

$$\mathbf{u} = \partial_t\mathbf{x} = \partial_t(\mathbf{x}-\mathbf{X}) = \partial_t\mathbf{w} , \qquad \partial_t\mathbf{u} = \partial_u^2\mathbf{w} , \tag{5.18}$$

where $\mathbf{w} = \mathbf{x} - \mathbf{X}$ is the elastic displacement.

Remember now that, in the linear elastic solid approximation, we can identify, on the right-hand side of (5.17), the Eulerian space coordinates with the Lagrangian coordinates. This means that we can take the operator $\nabla = \mathbf{e}_\alpha\partial_\alpha$ in the space variables of the undeformed body, i.e., in the Lagrangian space variables.

The governing equation (5.7) gives

$$\operatorname{div}\boldsymbol{\sigma} = 2\mu\operatorname{div}\mathbf{D} + \lambda\operatorname{div}[(\operatorname{trace}\mathbf{D})\mathbf{I}] . \tag{5.19}$$

According to formula (5.4) and elementary formulae of vector analysis (which the reader can easily reproduce if needed in coordinates) we have

$$\operatorname{div} \mathbf{D} = \tfrac{1}{2}\nabla(\operatorname{grad} \mathbf{w} + \operatorname{grad} \mathbf{w}^{\mathrm{T}}) = \tfrac{1}{2}(\operatorname{grad} \operatorname{div} \mathbf{w} + \tfrac{1}{2}\Delta \mathbf{w}),$$

$$\operatorname{div}((\operatorname{trace} \mathbf{D})\mathbf{I}) = \operatorname{div}(\operatorname{div} \mathbf{w}\mathbf{I}) = \mathbf{I}\operatorname{grad} \operatorname{div} \mathbf{w} = \operatorname{grad} \operatorname{div} \mathbf{w}.$$

Therefore substituting (5.18) and (5.19) into the momentum balance equation (5.17) we obtain the basic equation of the "dynamic" theory of elasticity:

$$\rho \partial_{tt}^2 \mathbf{w} = \mu \Delta \mathbf{w} + \frac{\mu}{1-2\nu} \operatorname{grad} \operatorname{div} \mathbf{w}. \tag{5.20}$$

Another form of this equation may also be useful. Again from elementary vector analysis we have

$$\operatorname{grad} \operatorname{div} \mathbf{w} = \Delta \mathbf{w} + \operatorname{curl} \operatorname{curl} \mathbf{w}. \tag{5.21}$$

Substituting this equation into (5.20) we reduce it to the form

$$\rho \partial_{tt}^2 \mathbf{w} = \frac{\mu}{1-2\nu} [2(1-\nu)\Delta \mathbf{w} + \operatorname{curl} \operatorname{curl} \mathbf{w}]. \tag{5.22}$$

In the static case, where there are no particle motions, the displacement vector $\mathbf{w}$ is time independent and the velocity is equal to zero; thus equation (5.20) assumes the form

$$(1-2\nu)\Delta \mathbf{w} + \operatorname{grad} \operatorname{div} \mathbf{w} = 0. \tag{5.23}$$

The basic equation (5.23) of the static theory of elasticity was published, in a different form, by the French scientist and engineer C. L. M. H. Navier (1827). (Note that this article was presented to the French Academy of Sciences in 1820 and read at its meeting in May 1821.) It has two important properties. Namely, according to this equation *the trace of the small-strain tensor* trace $\mathbf{D} = \operatorname{div} \mathbf{w}$ *is a harmonic function*. Indeed, let us perform the operation div over both parts of equation (5.23). The operations div and Δ are commutative, and so we obtain

$$\Delta \operatorname{div} \mathbf{w} = 0. \tag{5.24}$$

Now perform the operation Δ over both parts of equation (5.23). The operations grad and Δ are also commutative, and so taking into account (5.24) we obtain

$$\Delta \Delta \mathbf{w} = 0. \tag{5.25}$$

The static elasticity equation (5.23) requires us to prescribe boundary conditions. The most frequently used are the following alternatives.

(1) At the boundary $\partial\Omega$ of the deformable body Ω (Figure 5.2) a distribution of displacements $\mathbf{w}$ is prescribed (corresponding to a rigid punch):

$$\mathbf{w} = \mathbf{f}(\mathbf{x}), \qquad \mathbf{x} \in \partial\Omega. \tag{5.26}$$

Here $\mathbf{f}(\mathbf{x})$ is a given vector function; recall that the notation $\mathbf{x} \in \partial\Omega$ means that the point $\mathbf{x}$ belongs to the boundary $\partial\Omega$ of the body Ω.

(2) At the boundary $\partial\Omega$, with outward unit normal vector $\mathbf{n}$, the distribution of the *traction vector* $\mathbf{p}$ is prescribed:

$$\mathbf{p} = \boldsymbol{\sigma}\,\mathbf{n} = \mathbf{f}(\mathbf{x}) , \qquad \mathbf{x} \in \partial\Omega . \tag{5.27}$$

In the basic constitutive equation (5.7) (or (5.9)) the stress $\boldsymbol{\sigma}$ is expressed via the space derivatives of the displacement vector $\mathbf{w}$. Therefore the boundary value problem in this case is reduced to the prescription on the boundary of the body of a linear combination of space derivatives of the displacement vector $\mathbf{w}$, in terms of which the basic equations (5.23) are formulated.

(3) At the boundary $\partial\Omega$ the normal component of the displacement vector is prescribed, as is the tangential component of the traction vector (this constitutes a mixed problem):

$$u_n = \mathbf{w}\cdot\mathbf{n} = f(\mathbf{x}) , \qquad \mathbf{p}_t = \mathbf{p} - (\mathbf{p}\cdot\mathbf{n})\mathbf{n} = \mathbf{g}(\mathbf{x}) , \qquad \mathbf{x} \in \partial\Omega . \tag{5.28}$$

Here f is a prescribed scalar function and $\mathbf{g}(\mathbf{x})$ is a prescribed vector function.

As we can see, all these boundary value problems are linear *if the boundary $\partial\Omega$ is known beforehand*[3] and the basic equation (5.23) is also linear.

For dynamic problems governed by equations (5.20) or (5.22), initial conditions at a certain moment $t = t_0$ should also be prescribed:

$$\mathbf{w}(\mathbf{x}, t_0) = \mathbf{F}(\mathbf{x}) , \qquad \partial_t\mathbf{w}(\mathbf{x}, t_0) = \mathbf{G}(\mathbf{x}) . \tag{5.29}$$

Here $\mathbf{F}(\mathbf{x})$ and $\mathbf{G}(\mathbf{x})$ are vector functions prescribed in Ω.

Among the dynamic problems of elasticity, elastic waves play a special role. Consider an infinite body. It is possible to represent the elastic displacement field $\mathbf{w}(\mathbf{x}, t)$ as a sum of two vector fields,

$$\mathbf{w} = \mathbf{w}_1(\mathbf{x}, t) + \mathbf{w}_2(\mathbf{x}, t) , \tag{5.30}$$

where the field $\mathbf{w}_1$ is solenoidal, i.e. its divergence is equal to zero, while the field $\mathbf{w}_2(\mathbf{x}, t)$ is irrotational, so that

$$\operatorname{div}\mathbf{w}_1 = 0 , \qquad \operatorname{curl}\mathbf{w}_2 = 0 .$$

Substituting (5.30) into (5.20) and in turn performing the operations curl and div, we obtain

$$\operatorname{curl}\left(\partial_{tt}^2\mathbf{w}_1 - \frac{\mu}{\rho}\Delta\mathbf{w}_1\right) = 0 ,$$

$$\operatorname{div}\left(\partial_{tt}^2\mathbf{w}_2 - \frac{2(1-\nu)\,\mu}{(1-2\nu)\,\rho}\Delta\mathbf{w}_2\right) = 0 . \tag{5.31}$$

---

[3] The importance of this reservation will become clear when we consider fracture problems.

However, by the definitions of $\mathbf{w}_1$ and $\mathbf{w}_2$ it follows that

$$\mathrm{div}\left(\partial_{tt}^2\mathbf{w}_1 - \frac{\mu}{\rho}\Delta\mathbf{w}_1\right) = 0,$$

$$\mathrm{curl}\left[\partial_{tt}^2\mathbf{w}_2 - \frac{2(1-\nu)\,\mu}{(1-2\nu)\,\rho}\Delta\mathbf{w}_2\right] = 0. \tag{5.32}$$

Remember that if $\mathrm{curl}\,A = 0$ for a vector $\mathbf{A}$ then there exists a scalar function ϕ such that $\mathbf{A} = \mathrm{grad}\,\phi$. If, in addition, $\mathrm{div}\,\mathbf{A} = 0$ then ϕ must be a harmonic function, so that $\Delta\phi = 0$. If the body is infinite, it follows that $\phi \equiv \mathrm{const}$, because any function that is harmonic in an entire space must be a constant. Assuming that at infinity the displacement vanishes we obtain that representation (5.30) is unique and, from equations (5.31) and (5.32), we find that in an elastic medium two types of waves are propagated, described by two wave equations with different wave speeds a_1, a_2:

$$\partial_{tt}^2\mathbf{w}_1 - a_1^2\Delta\mathbf{w}_1 = 0, \qquad a_1^2 = \left(\frac{\mu}{\rho}\right),$$

$$\partial_{tt}^2\mathbf{w}_2 - a_2^2\Delta\mathbf{w}_2 = 0, \qquad a_2^2 = \frac{2(1-\nu)\,\mu}{(1-2\nu)\,\rho} > a_1^2. \tag{5.33}$$

In bounded elastic bodies these two types of wave do not propagate independently, because at the boundaries a wave of one type produces, generally speaking, waves of the other type.

In waves of the first type, all that takes place is a distortion of the particles: there is no change of volume because $\mathrm{div}\,\mathbf{w}_1 = 0$. Therefore these waves are called *shear waves*. In distinction to these, irrotational waves of the second type produce volume changes. Therefore these waves are called *compression–expansion waves*.

5.3 Plane problem in the theory of elasticity

In the previous section we saw that the basic boundary value problems of the theory of elasticity are linear. For more than a century, most of the mathematical theory of elasticity consisted of developing effective methods for solving wide classes of problems. In recent decades, however, computational methods have become increasingly important. There are many excellent comprehensive treatises, textbooks and monographs where these methods are explored. In the present book we do not claim to consider the available methods of solution for elasticity problems in any detail: these are generally beyond our scope. However, to demonstrate some principal properties of the mathematical models of solids we will need some explicit analytic solutions of them. Using such solutions we will be able to understand important qualitative properties of the models and also their singularities, which are of special interest because they demonstrate what happens when the model breaks down and what we will need from more general models.

Therefore we choose for more detailed consideration here a very efficient method which is applicable in the case of *plane strains*. This term corresponds to the case when the component of the displacement vector $\mathbf{w}$ along a certain axis x_3 of an orthonormal Cartesian coordinate system is identically equal to zero and the two other components depend only on the coordinates x_1 and x_2 orthogonal to x_3:

$$w_1 = w_1(x_1, x_2), \qquad w_2 = w_2(x_1, x_2), \qquad w_3 = 0. \tag{5.34}$$

Of course, plane strain is a very special case of deformation. It occurs in some interesting practical problems, for instance in the design of aircraft wings or wide dams. However, in this book this method will be used mainly to reveal some general intermediate-asymptotic properties of elastic fields and in particular to present the phenomenon of fracture.

The method that we present here uses the theory of functions of a complex variable. It is now an everyday tool in technical design, even when the elastic fields involved are not plane strain fields. The invention of this method is considered to be one of the fundamental twentieth century achievements in the theory of elasticity; its development is due mainly to the Soviet scientists G. V. Kolosov (1909) and N. I. Muskhelishvili (1963).[4]

In the plane strain case, Hooke's law for the components σ_{11} and σ_{22} of the stress tensor (see equation (5.10)) can be written as

$$
\begin{aligned}
\sigma_{11} &= \frac{E}{(1+\nu)(1-2\nu)} \left[(1-\nu)\partial_1 w_1 + \nu \partial_2 w_2 \right], \\
\sigma_{22} &= \frac{E}{(1+\nu)(1-2\nu)} \left[\nu \partial_1 w_1 + (1-\nu)\partial_2 w_2 \right];
\end{aligned}
\tag{5.35}
$$

remember that w_3 is equal to zero and that w_1, w_2 do not depend on x_3. Adding these equations we obtain

$$\sigma_{11} + \sigma_{22} = \frac{E}{(1+\nu)(1-2\nu)} \left(\partial_1 w_1 + \partial_2 w_2 \right).$$

However, according to (5.24) and bearing in mind that $\partial_3 w_3 = 0$ and $\partial_\alpha w_\alpha = \operatorname{div} \mathbf{w}$ is a harmonic function, we find that $\sigma_{11} + \sigma_{22}$ is also a harmonic function:

$$\Delta(\sigma_{11} + \sigma_{22}) = 0.$$

---

[4] Here there is a certain mystery. When S. A. Chaplyguine, the great Russian applied mathematician, died in 1942, Professor L. N. Sretensky was commissioned to consider his remaining manuscripts. He found among them a manuscript containing the basic ideas and results of this method that was definitely written earlier than the first fundamental memoir of Kolosov (1909).

In the static equilibrium case the momentum balance equation assumes the form $\operatorname{div}\boldsymbol{\sigma} = 0$. The quantity $\operatorname{div}\boldsymbol{\sigma}$ is a vector, and when projected onto the x_1 and x_2 axes this equation takes the form

$$\begin{aligned}\partial_1\sigma_{11} + \partial_2\sigma_{12} &= 0\,,\\ \partial_1\sigma_{21} + \partial_2\sigma_{22} &= 0\,.\end{aligned} \tag{5.36}$$

Remember that according to Hooke's law the stress components do not depend on x_3 and that $\sigma_{12} = \sigma_{21}$.

Now recall our discussion of the plane motion of an ideal incompressible fluid. The continuity equation

$$\partial_1 u_1 + \partial_2 u_2 = 0$$

(u_1, u_2 are velocity components) allowed us to introduce the stream function ψ so that $u_1 = \partial_2\psi$, $u_2 = -\partial_1\psi$ constitute the condition for integrability. In a completely analogous way the equations of elastic equilibrium (5.36) can be used as an integrability condition allowing us to express all the non-vanishing components of the stress tensor $\sigma_{11}, \sigma_{22}, \sigma_{12}$ via a single function, namely the *stress function* $\theta(x_1, x_2)$:

$$\begin{aligned}\sigma_{11} &= \partial_{22}^2\theta\,,\\ \sigma_{22} &= \partial_{11}^2\theta\,,\\ \sigma_{12} = \sigma_{21} &= -\partial_{12}^2\theta\,.\end{aligned} \tag{5.37}$$

Indeed, the substitutions (5.37) satisfy the equations of elastic equilibrium (5.36) identically. The stress function was introduced by the British astronomer Airy and so is called the Airy function. According to (5.37)

$$\sigma_{11} + \sigma_{22} = \Delta\theta\,,$$

and, thanks to the harmonicity of the function $\sigma_{11} + \sigma_{22}$, we obtain that the function θ is *biharmonic*:

$$\Delta\,\Delta\theta = 0\,. \tag{5.38}$$

The argument presented here is close to that used when we proved the harmonicity of the potential and stream functions for plane potential flows of an ideal incompressible fluid. The general solution to the Laplace equation $\Delta\phi = 0$ in the plane case is $\phi = \operatorname{Re} w(z)$, where $w(z)$ is an arbitrary analytic function of a complex variable $z = x_1 + ix_2$. The general solution of the biharmonic equation, obtained by the French mathematician E. Goursat, is constructed in a way that is only slightly more complicated. It is expressed via not one but two analytic functions of a complex variable z. Indeed, it follows from (5.38) that $\Delta\theta = P$ is a harmonic function. Therefore $P + iQ = f(z)$ is an analytic function of a complex variable z, where Q

is a function of two variables x_1 and x_2 *harmonically conjugate* to P; the functions P and Q are related by the Cauchy–Riemann equations

$$\partial_1 P = \partial_2 Q , \qquad \partial_2 P = -\partial_1 Q . \tag{5.39}$$

Introduce an analytic function

$$\phi(z) = p + iq = \tfrac{1}{4} \int f(z)\, dz \tag{5.40}$$

so that $\phi' = \tfrac{1}{4} f(z) = \tfrac{1}{4}(P + iQ)$. The harmonically conjugate functions $p(x_1, x_2)$, $q(x_1, x_2)$ also satisfy the Cauchy–Riemann equations as well as the equations

$$\partial_1 p = \partial_2 q = \tfrac{1}{4} P , \qquad \partial_2 p = -\partial_1 q = -\tfrac{1}{4} Q . \tag{5.41}$$

We observe easily that $\Delta(p x_1) = 2\partial_1 p$ and $\Delta(q x_2) = 2\partial_2 q = 2\partial_1 p$ and therefore

$$\Delta(\theta - p x_1 - q x_2) = 0 , \tag{5.42}$$

so that the harmonic function $\theta - p x_1 - q x_2 = p_1$ is also the real part of an analytic function $\chi(z) = p_1 + iq_1$ of a complex variable z. But $p x_1 + q x_2$ is evidently the real part of a function $\bar{z}\phi(z) = (p + iq)(x_1 - ix_2)$. Therefore a general solution to the biharmonic equation (5.38) can be represented in the form

$$\theta = \mathrm{Re}\left[\bar{z}\phi(z) + \chi(z)\right] . \tag{5.43}$$

Here $\phi(z), \chi(z)$ are arbitrary analytic functions of the complex variable $z = x_1 + ix_2$.

The problem is therefore to find these functions ϕ and χ for the elastic field in which we are interested. However, the boundary conditions, as we have seen, are presented via the displacement and stress. Therefore we now have to express the stress and displacement components via these analytic functions. According to the relations (5.37), which introduced the stress function, it is possible to rewrite Hooke's law (5.35) in the plane strain case in the form

$$
\begin{aligned}
\partial_{22}^2 \theta &= \frac{E}{(1 + \nu)(1 - 2\nu)} \left[(1 - \nu)\partial_1 w_1 + \nu \partial_2 w_2 \right] , \\
\partial_{11}^2 \theta &= \frac{E}{(1 + \nu)(1 - 2\nu)} \left[\nu \partial_1 w_1 + (1 - \nu)\partial_2 w_2 \right] .
\end{aligned}
\tag{5.44}
$$

We can solve these equations for $\partial_1 w_1$ and $\partial_2 w_2$:

$$
\begin{aligned}
\partial_1 w_1 &= \frac{1 + \nu}{E} \left(\partial_{22}^2 \theta - \nu \Delta\theta \right) , \\
\partial_2 w_2 &= \frac{1 + \nu}{E} \left(\partial_{11}^2 \theta - \nu \Delta\theta \right) .
\end{aligned}
\tag{5.45}
$$

Transforming the first, we find

$$\frac{E}{1+\nu}\,\partial_1 w_1 = \partial_{22}^2\theta - \nu\Delta\theta$$
$$= (1-\nu)\Delta\theta - \partial_{11}^2\theta$$
$$= -\partial_{11}^2\theta + 4(1-\nu)\partial_1 p\,, \tag{5.46}$$

because, according to our previous arguments, $\Delta\theta = P = 4\partial_1 p$. Integrating (5.46) over the variable x_1, we obtain

$$\frac{E}{1+\nu}\,w_1 = -\partial_1\theta + 4(1-\nu)p + f_1(x_2)\,, \tag{5.47}$$

where $f_1(x_2)$ is an arbitrary function of x_2. In an analogous way we obtain from the second equation in (5.45)

$$\frac{E}{1+\nu}\,w_2 = -\partial_2\theta + 4(1-\nu)q + f_2(x_1)\,, \tag{5.48}$$

where f_2 is also an arbitrary function but of x_1. Note now that for the stress component $\sigma_{12} = -\partial_{12}^2\theta$ the following relation is valid:

$$\sigma_{12} = -\partial_{12}^2\theta = \frac{E}{2(1+\nu)}\,(\partial_1 w_2 + \partial_2 w_1)\,; \tag{5.49}$$

this also follows from Hooke's law for the plane-strain case. So, differentiating (5.47) by x_2, and (5.48) by x_1 and substituting into (5.49), we obtain that $f_1'(x_2) + f_2'(x_1) = 0$. The functions f_1 and f_2 are functions of different arguments, therefore $f_1'(x_2) = \alpha$, $f_2'(x_1) = -\alpha$ and

$$f_1 = \alpha x_2 + \beta\,, \qquad f_2 = -\alpha x_1 + \gamma\,,$$

where α, β, γ are constants. Obviously these functions are the x_1, x_2 components of a solid body displacement. If the reference system of the observer is attached to the elastic body under consideration then these functions can be taken as equal to zero. We obtain

$$\frac{E}{1+\nu}\,(w_1 + iw_2) = -(\partial_1\theta + i\partial_2\theta) + 4(1-\nu)(p+iq)$$
$$= -(\partial_1\theta + i\partial_2\theta) + 4(1-\nu)\phi(z)\,. \tag{5.50}$$

According to (5.43) we have

$$2\theta = \bar{z}\phi + z\bar{\phi} + \chi + \bar{\chi}\,. \tag{5.51}$$

We find by differentiating this relation that

$$2\partial_1\theta = \bar{z}\phi' + \phi + z\overline{\phi'} + \bar{\phi} + \chi' + \overline{\chi'}\,,$$
$$2\partial_2\theta = i(-\phi + \bar{\phi} + \bar{z}\phi' - z\overline{\phi'} + \chi' + \overline{\chi'})\,, \tag{5.52}$$

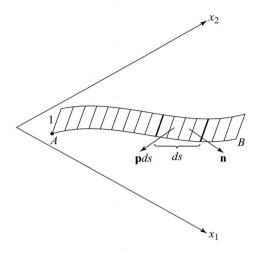

Figure 5.4 Representation of the traction **p** via Kolosov–Muskhelishvili potentials; see the text.

and it is easy to find the following relation for the important quantity $\partial_1\theta + i\partial_2\theta$:

$$\partial_1\theta + i\partial_2\theta = \phi + z\overline{\phi'} + \bar{\psi} . \tag{5.53}$$

Here we have introduced the notation $\chi'(z) = \psi(z)$. Now we substitute (5.53) into (5.50) and ultimately obtain the formula expressing the displacement in terms of the functions ϕ and ψ. In the literature it is known as the *first Kolosov–Muskhelishvili formula*,

$$\frac{E}{1+v}(w_1 + iw_2) = \kappa\phi - z\overline{\phi'} - \bar{\psi} . \tag{5.54}$$

Here the notation $\kappa = 3 - 4v$ has been used.

Now we will obtain analogous formulae for the stress components. Let us take a cylindrical surface of unit height supported by a contour AB with external normal **n** (Figure 5.4). An elementary force **p** ds acts on the element of this cylindrical surface supported by a line element ds of the contour AB. Here obviously (see formulae (1.32), (1.33)) the components of the traction force **p** are

$$p_1 = \sigma_{11}n_1 + \sigma_{12}n_2 , \qquad p_2 = \sigma_{12}n_1 + \sigma_{22}n_2 , \tag{5.55}$$

where n_1, n_2 are the projections of the unit normal vector **n**, equal to respectively dx_2/ds, $-dx_1/ds$. We know how the stress components are expressed in terms of the stress function θ:

$$\begin{aligned} p_1 &= \partial_{22}^2\theta\frac{dx_2}{ds} + \partial_{12}^2\theta\frac{dx_1}{ds} = \frac{d}{ds}(\partial_2\theta) , \\ p_2 &= -\frac{d}{ds}(\partial_1\theta) , \end{aligned} \tag{5.56}$$

and from this we obtain an important general relation,

$$p_1 + ip_2 = \frac{d}{ds}(\partial_2\theta - i\partial_1\theta) = -i\frac{d}{ds}(\partial_1\theta + i\partial_2\theta)\,. \tag{5.57}$$

Now take the direction ds to lie along the axis x_2, so that the vector $\mathbf{n}$ is directed along the axis x_1: we can choose the line AB in an arbitrary way. In this case, from (5.57) we get

$$
\begin{aligned}
\sigma_{11} + i\sigma_{12} &= -i\partial_2(\partial_1\theta + i\partial_2\theta) \\
&= -i\partial_2(\phi + z\overline{\phi'} + \bar{\psi}) \\
&= \phi' + \overline{\phi'} - z\overline{\phi''} - \overline{\psi'}\,.
\end{aligned}
\tag{5.58}
$$

Now let us direct ds along the axis x_1, so that the direction of $\mathbf{n}$ is opposite to that of the axis x_2. We obtain from (5.57)

$$-\sigma_{12} - i\sigma_{22} = -i(\phi' + \overline{\phi'} + z\overline{\phi''} + \overline{\psi'})\,,$$

so that

$$\sigma_{22} - i\sigma_{12} = \phi' + \overline{\phi'} + z\overline{\phi''} + \overline{\psi'}\,. \tag{5.59}$$

Formulae (5.58) and (5.59) ultimately allow us to obtain two relations in which the components of the stress tensor are represented in terms of the functions ϕ and ψ; these relations are the *second and third Kolosov–Muskhelishvili formulae*,

$$
\begin{aligned}
\sigma_{11} + \sigma_{12} &= 4\,\mathrm{Re}\,\Phi(z)\,, \\
\sigma_{22} - \sigma_{11} + 2i\sigma_{12} &= 2(\bar{z}\Phi' + \Psi)\,.
\end{aligned}
\tag{5.60}
$$

The analytic functions $\Phi = \phi'$ and $\Psi = \psi'$ are called the *Kolosov–Muskhelishvili potentials*.

5.4 Analytical solutions of some special problems in plane elasticity

In this section we will present solutions of two problems of special significance.

5.4.1 A concentrated load in an infinite elastic body

A concentrated load of magnitude $\mathbf{X}(X_1, X_2)$ per unit length in an infinite elastic body is uniformly distributed along the x_3 axis.[5] The problem is linear, but the arguments x_1 and x_2 enter in different ways because, in the complex variable Y_z, x_2 enters with a factor i. Therefore we will consider separately the cases $X_2 = 0$ and $X_1 = 0$.

---

[5] Such a loading is easy to imagine as applied to a thin rigid rod soldered into a thin cylindrical hole drilled along its x_3 axis.

In the first case the Kolosov–Muskhelishvili potentials depend only on the quantities X_1 and z and on a dimensionless property of the elastic body, its Poisson ratio ν. The Young modulus is not relevant because only the loads, not the displacements, enter the boundary condition and, according to formula (5.60), the potentials Φ and Ψ are not influenced by E. The dimensions of the quantities that we need to consider are evidently as follows: $[\Phi] = [\Psi] = FL^{-2}$ is the dimension of stress, $[X_1] = FL^{-1}$ is the dimension of force distributed along a line and $[z] = L$, $[\nu] = 1$. Here F is the dimension of force and L is the dimension of length. Dimensional analysis gives manifestly

$$\Phi = \frac{X_1}{z} A(\nu) , \qquad \Psi = \frac{X_1}{z} B(\nu) . \tag{5.61}$$

Here $A = A_1 + iA_2$, $B = B_1 + iB_2$ are complex constants. It follows from (5.61) that

$$\phi = X_1 A(\nu) \ln z , \qquad \psi = X_1 B(\nu) \ln z . \tag{5.62}$$

The constants A_1, A_2, B_1, B_2 are determined in the following way. First, the complex displacement $w_1 + iw_2$ must be a single-valued function: moving around the origin along a closed contour should return us to the same value of displacement. Put $z = re^{i\omega}$ in (5.62), and substitute the result into the first Kolosov–Muskhelishvili formula (5.54). It is easy to show that the polar angle ω will enter the right-hand side of (5.54) with coefficient

$$X_1[-\kappa A_2 + B_2 + i(\kappa A_1 + B_1)] . \tag{5.63}$$

The real and imaginary parts of this expression must each vanish because the other terms on the right-hand side of (5.54) are single-valued functions. Furthermore, for the bulk force acting on the cylindrical surface of unit height supported by the contour AB, formula (5.57) yields the expression

$$\int_A^B (p_1 + ip_2) \, ds = -i(\partial_1\theta + i\partial_2\theta) \,|_A^B = -i(\phi + z\overline{\phi'} + \bar{\psi}) \,|_A^B . \tag{5.64}$$

If the contour AB is closed, so that points A and B coincide, the bulk force is equal to the increment in the function $-i(\phi + z\overline{\phi'} + \bar{\psi})$ after going around the origin. Using (5.62) it is easy to see this increment is equal to

$$2\pi X_1[A_1 - B_1 + i(A_2 + B_2)] . \tag{5.65}$$

However, for the equilibrium of a cylindrical body supported by a closed contour around the origin at which the concentrated force is applied, the bulk force acting on the cylindrical surface from the ambient part of the body is equal to $-X_1$. From that condition and from the vanishing of the real and imaginary parts of expression (5.63), a set of equations for coefficients A_1, A_2, B_1, B_2 is obtained:

$$\kappa A_1 + B_1 = 0, \quad -\kappa A_2 + B_2 = 0, \quad 2\pi(A_1 - B_1) = -1, \quad A_2 + B_2 = 0 . \quad (5.66)$$

Solving these equations we eventually find the following relations for the Kolosov–Muskhelishvili potentials:

$$\Phi = -\frac{X_1}{2\pi(1+\kappa)z}, \qquad \Psi = \frac{\kappa X_1}{2\pi(1+\kappa)z} . \qquad (5.67)$$

The case $X_1 = 0$ can be considered in a completely analogous way, and the following expressions for Φ and Ψ are obtained:

$$\Phi = -\frac{iX_2}{2\pi(1+\kappa)z}, \qquad \Psi = \frac{i\kappa X_2}{2\pi(1+\kappa)z} . \qquad (5.68)$$

Thus in the general case the Kolosov–Muskhelishvili potentials for a concentrated force applied along the x_3 axis have the following form:

$$\begin{aligned}
\Phi &= -\frac{X_1 + iX_2}{2\pi(1+\kappa)z} = -\frac{X_1 + iX_2}{8\pi(1-\nu)z}, \\
\Psi &= \frac{\kappa(X_1 - iX_2)}{2\pi(1+\kappa)z} = \frac{\kappa(X_1 - iX_2)}{8\pi(1-\nu)z} .
\end{aligned} \qquad (5.69)$$

The solutions obtained for a concentrated force allow the construction of an important solution for a so-called force dipole. Indeed, take a concentrated force 0, X_2 at the point $z_0 = -ih$ and a force equal in magnitude but oppositely directed, 0, $-X_2$, at the point $-z_0$. Then assume that h becomes infinitesimally small, but in such a way that the product $X_2 h = T$ remains constant. In the limit we obtain, as it is easy to show, the following expressions for the Kolosov–Muskhelishvili potentials:

$$\begin{aligned}
\Phi &= \frac{T}{\pi(1+\kappa)z^2} = \frac{T}{4\pi(1-\nu)z^2}, \\
\Psi &= \frac{\kappa T}{\pi(1+\kappa)z^2} = \frac{\kappa T}{4\pi(1-\nu)z^2} .
\end{aligned} \qquad (5.70)$$

We emphasize, and it will be needed later, that the Kolosov–Muskhelishvili potentials for a force dipole decay at infinity as z^{-2}.

5.4.2 Infinitely thin plane cut in an infinite elastic body

Consider an infinitely thin cut along the x_1 axis from $x_1 = -\ell$ to $x_1 = \ell$ in an infinite elastic body under the action of a system of loads that is symmetric with

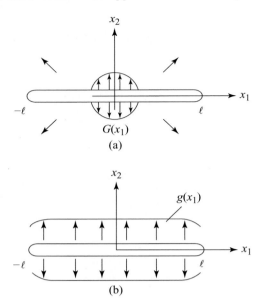

Figure 5.5 Decomposition of the loads acting on a body with a thin cut.

respect to the x_1 and x_2 axes (Figure 5.5). In this case we can clearly represent the elastic field as a sum of two fields. The first elastic field corresponds to an infinite body without a cut, under the action of the same loads. Clearly at the position of the cut there will appear a distribution of normal stress $\sigma_{22} = g(x_1)$. Owing to the symmetry of the loads the tangential stress at the position of the cut should vanish: $\sigma_{12} = 0$. Indeed, if there were a non-zero tangential stress and we turned the loaded body over then the tangential stress would change its sign whilst owing to their symmetry the loads would remain the same.

The second field corresponds to an infinite body with the cut and with symmetric loads applied but only at the surface of the cut. These loads must compensate the loads $\sigma_{22} = g(x_1)$ appearing in the plane of the cut and also take into account the normal loads $\sigma_{22} = G(x_1)$ applied directly of the cut side. Thus, in the second field the normal loads at the cut sides are distributed according to

$$\sigma_{22} = -g(x_1) + G(x_1) \, .$$

The first field can be obtained by summing the solutions given in the previous section. It is, in principle, an elementary task, so the function $g(x_1)$ can be considered to be known. The function $G(x_1)$ is prescribed as part of the boundary conditions in the basic problem.

To obtain the second field we use the second and third Kolosov–Muskhelishvili formulae (5.60). According to the third formula, at the cut and its continuation along the x_1 axis, where obviously $\bar{z} = z$, we have

$$\sigma_{22} - \sigma_{11} + 2i\sigma_{12} = 2(z\Phi' + \Psi) \,. \tag{5.71}$$

We emphasize that on the right-hand side we have an analytic function $z\Phi' + \Psi$ of a complex variable, because in (5.60) we replaced $\bar{z}$ by z. But at the cut and its continuation the shear stress vanishes, $\sigma_{12} = 0$; therefore the imaginary part of this function also vanishes:

$$\text{Im} \, (z\Phi' + \Psi) = 0 \,. \tag{5.72}$$

Evidently the Kolosov–Muskhelishvili potentials behave at infinity like those for a force dipole, i.e. they decay as $1/z^2$. Therefore the analytic function $z\Phi' + \Psi$ must similarly decay at infinity. But this means that the function $z\Phi' + \Psi$ is identically equal to zero: this is an elementary fact from the theory of analytic functions. Thus

$$\Psi = -z\Phi' \,. \tag{5.73}$$

We find by integrating (5.73) that

$$\psi = -\int z\Phi' dz = -z\phi' + \phi \,. \tag{5.74}$$

Thus the problem under consideration is reduced to the determination of a single analytic function Φ. Now substitute into the third Kolosov–Muskhelishvili formula (see (5.60)) relation (5.73) for Ψ; for the problem under consideration the formula reduces to

$$\sigma_{22} - \sigma_{11} + 2i\sigma_{12} = 2(\bar{z} - z)\Phi' \,. \tag{5.75}$$

In particular, from (5.75) it follows that at the cut, and also at its continuation, $\sigma_{22} = \sigma_{11}$. However, according to the second Kolosov–Muskhelishvili formula, $\sigma_{11} + \sigma_{22} = 4 \, \text{Re} \, \Phi$. Therefore at the cut $-\ell \leq x_1 \leq \ell$ we have

$$\text{Re} \, \Phi(z) = \tfrac{1}{2}[-g(x_1) + G(x_1)] \,. \tag{5.76}$$

This condition, together with the condition $\Phi(z) \to 0$ at $z \to \infty$, determines the solution of the Dirichlet problem in the theory of harmonic functions for $\text{Re} \, \Phi(z)$. The standard solution of this problem, which can be readily found in any textbook on complex analysis, gives

$$\Phi(z) = -\frac{1}{2\pi \sqrt{z^2 - \ell^2}} \int_{-\ell}^{\ell} \frac{[g(\xi) - G(\xi)] \sqrt{\ell^2 - \xi^2} \, d\xi}{\xi - z} \,. \tag{5.77}$$

Formally, this expression gives us the final solution to the problem under consideration. The stress field is obtained by further application of the Kolosov–Muskhelishvili formulae (5.60). The displacement field is obtained by the application of the first Kolosov–Muskhelishvili formula, (5.54), which, according to (5.74), assumes the form

$$\frac{E}{1+\nu}(w_1 + iw_2) = (3 - 4\nu)\phi + (\bar{z} - z)\overline{\phi'(z)} - \phi(z) . \tag{5.78}$$

Investigation of the obtained solution and its singular properties will be presented in the next chapter.

6

The linear elastic solid approximation. Applications: brittle and quasi-brittle fracture; strength of structures

6.1 The problem of structural integrity

The science of studying the strength of structures (i.e. structural integrity), traditionally known as the strength of materials (this term, as we will see, is not quite adequate) appeared like Athena from Zeus' head. It was created by Galileo Galilei, and was presented in his book *Dialogues Concerning Two New Sciences* (Galilei, 1638), which appeared, by the way, not in his home country, Italy, but in Leiden, The Netherlands. He was able to overcome the Catholic Church's prohibition to publish anything by sending parts of his manuscript to Leiden via his friends.[1] Galilei gave the following definition of the subject of this new science: "New science, treating the resistance which solid bodies offer to fracture." It is remarkable that now, nearly 400 years since its publication, this definition sounds quite modern. Roughly speaking, the problem is to determine the limiting load which a *structure* is able to carry. The goal of the tensile test, shown in Figure 5.1 and taken from his book (Galilei, 1638), was to determine the limiting load.

Robert Hooke's fundamental paper published 40 years later, in 1678, which as we saw in the previous chapter laid the foundation of the theory of linear elasticity, marked, strictly speaking, a deviation from the path formulated by Galilei. In fact, this paper founded the science treating the *deformation* of elastic solid bodies due to the action of loads applied to them. This is a remarkable, very important, but different problem.

In particular, the theory of linear elasticity cannot yield a model of the fracture phenomenon. Indeed, having a solution $\mathbf{w}(\mathbf{x})$ of a problem of the theory of elasticity for a certain load P, we obtain the solution of the problem for an arbitrarily large load AP simply by multiplying the solution $\mathbf{w}(\mathbf{x})$ by the factor A. So, according to the linear elastic model of a body, the structure can carry arbitrarily large loads,

---

[1] Galilei was rehabilitated by the Church quite recently, in 1992, in the pontificate of John Paul II.

and of course this contradicts everyday experience. Furthermore, fracture is not necessarily accompanied by irreversible, e.g. plastic, deformation: brittle or quasi-brittle (see below) fracture is often the case. (Remember the porcelain cup that you broke in childhood: its parts can be assembled back together so that the fracture is not noticeable, but even so punishment was unavoidable!)

So, the development of the Galilean science of the Strength of Structures remained to a large extent frozen for more than 250 years, whereas the mathematical theory of elasticity flourished.

In the mid nineteenth century the problem of Strength attracted the attention of genius, James Clerk Maxwell, who touched upon it in his letter to William Thompson (the future Lord Kelvin).[2] Maxwell made the following instructive statement: "When the strain energy of distortion reaches a certain limit, then the element will begin to give way." He emphasized further, "This is the first time that I have put pen to paper on this subject. I have never seen any investigation of the question: *Given the mechanical strain in three directions on an element, when will it give way?*" Maxwell's idea became known only after publication of his letter. After a certain time Maxwell's idea found its expression in the Huber–von Mises–Hencky criterion of fracture, which can be formulated as follows:

The fracture of a structure is a local event, and it occurs when for a certain element of a structure the positively defined quantity, proportional to the distortion energy,

$$(\sigma_1 - \sigma_2)^2 + (\sigma_2 - \sigma_3)^2 + (\sigma_3 - \sigma_1)^2 \qquad (6.1)$$

(where σ_1, σ_2, σ_3 are the principal stresses) reaches a certain limit $2k^2$.

This limit, if one believes this criterion, can be obtained from a one-dimensional tensile test, for which $\sigma_1 = \sigma_Y$ or σ_f and $\sigma_2 = \sigma_3 = 0$ (σ_Y is the yield stress, σ_f is the fracture (ultimate) stress); $k^2 = \sigma_Y^2$ or $k^2 = \sigma_f^2$.

The Huber–von Mises–Hencky criterion was and, up to the present time, is in wide use for engineering evaluations for the strength of structures.

The subsequent consideration of brittle fracture, where the material remains ideally elastic up to the fracture, showed that this criterion is not quite correct. *Fracture is not a local but a global event, and the condition for the fracture of structures cannot be obtained using a local criterion similar to (6.1).*

6.2 Defects and cracks

A crucially important step in the development of the science of the strength of structures was performed by the British scientist and engineer A. A. Griffith, a

---

[2] See the classic text of S. P. Timoshenko (1953).

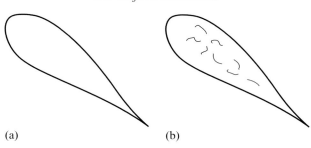

(a)   (b)

Figure 6.1 (a) The ideal shape of a structure as it appears in a designer's drawing. (b) The real shape of a structure where defects and cracks are also presented.

disciple of G. I. Taylor. Following him, the statement of the problem of the strength of structures can be presented as follows. The idealized shape of a structure, as presented in the drawing of a designer (Figure 6.1a), is incomplete. In fact, the situation regarding boundaries is more complicated. In addition to the "legal" boundaries the structure always has "illegal" ones (Figure 6.1b), the boundaries of the defects and, particularly, cracks which are always unavoidable in real structures. Their surfaces are part of the structure's boundary, and this part should also be taken into account.

Seemingly, but only seemingly, treating the material as ideally elastic changes nothing, at least in the case of brittle fracture. Indeed, in principle the linear theory of elasticity allows us to obtain the solution for an elastic field, i.e. the distribution of displacements, stresses and strains for multiconnected bodies having an arbitrary boundary shape. In particular, the theory of elasticity allows the construction of such solutions for bodies with cracks, represented as infinitely thin cuts, as demonstrated in Section 5.4.2.

Here, however, two substantial complications should be taken into account. First, in contrast with the "legal" boundaries, the shapes and sizes of cracks do not remain invariant under the action of loads. Therefore in prescribing the loads acting on a structure, the sizes and shapes of the cracks cannot be prescribed beforehand as in the case of "legal" boundaries. The second complication is that, generally speaking, the solutions to elasticity problems for bodies with infinitely thin cuts have singularities at the tips of cuts which make these solutions inappropriate as intermediate asymptotics.

To demonstrate this, we return to Section 5.4.2, an infinite body with a plane cut under the action of symmetric loads. For this example, the displacements and stresses are determined (see Section 5.4) by a single Kolosov–Muskhelishvili potential $\Phi(z) = \phi'(z)$:

$$\frac{E}{1+\nu}(w_1 + iw_2) = (3 - 4\nu)\phi + (\bar{z} - z)\overline{\phi'} - \overline{\phi(z)}, \tag{6.2}$$

$$\sigma_{11} + \sigma_{22} = 4\,\text{Re}\,\Phi(z), \tag{6.3}$$

$$\sigma_{22} - \sigma_{11} + 2i\sigma_{12} = 2(\bar{z} - z)\Phi'(z).$$

For the potential $\Phi(z)$ we obtain the expression

$$\Phi(z) = -\frac{1}{2\pi\sqrt{z^2 - \ell^2}} \int_{-\ell}^{\ell} \frac{g(\xi)\sqrt{\ell^2 - \xi^2}\,d\xi}{\xi - z}. \tag{6.4}$$

We remind the reader that $g(\xi)$ is the distribution of the normal stress σ_{22} in the uncut body on the plane of the cut. It can be obtained as a combination of the solutions in Section 5.4.1.

Let us investigate the elastic field in a small vicinity of the tip of the cut $z = \ell$. We put $z = \ell + s$, where $|s| \ll \ell$ and s is a real number that is positive for a straight continuation of the cut and negative on the cut itself.

At the cut, and its continuation, we have $\bar{z} = z$; therefore, according to (6.3), $\sigma_{11} = \sigma_{22} = 2\,\text{Re}\,\Phi(z)$, $z^2 - \ell^2 = s(2\ell + s)$ and finally

$$(\sigma_{22})_{x_2=0,x_1=\ell+s} = -\frac{1}{\pi\sqrt{s(2\ell + s)}} \int_{-\ell}^{\ell} \frac{g(\xi)\sqrt{\ell^2 - \xi^2}\,d\xi}{\xi - \ell - s}. \tag{6.5}$$

For small s, $(2\ell + s)^{-1/2} = [2\ell(1 + s/2\ell)]^{-1/2} \simeq (1/\sqrt{2\ell})(1 - s/4\ell)$. To avoid dealing with divergent integrals we represent the right-hand side of (6.5) in the form

$$-\frac{1}{\pi\sqrt{s(2\ell)}}\left(1 - \frac{s}{4\ell}\right)\left[\int_{-\ell}^{\ell} g(\xi)\sqrt{\ell^2 - \xi^2}\left(\frac{1}{\xi - \ell} + \frac{1}{\xi - \ell - s} - \frac{1}{\xi - \ell}\right)d\xi\right]$$

$$= \frac{1}{\pi\sqrt{s(2\ell)}}\left(1 - \frac{s}{4\ell}\right)\left[\int_{-\ell}^{\ell} g(\xi)\sqrt{\frac{\ell + \xi}{\ell - \xi}}\,d\xi + I_1 + I_2\right],$$

where

$$I_1 = -\int_{-\ell}^{\ell} [g(\xi) - g(\ell)]\sqrt{\ell^2 - \xi^2}\left(\frac{1}{\xi - \ell - s} - \frac{1}{\xi - \ell}\right)d\xi,$$

$$I_2 = -g(\ell)\int_{-\ell}^{\ell} \sqrt{\ell^2 - \xi^2}\left(\frac{1}{\xi - \ell - s} - \frac{1}{\xi - \ell}\right)d\xi.$$

It is now easy to show that $I_1 = O(s)$, $I_2 = g(\ell)(2\ell)^{1/2}\pi\sqrt{s}$, so that

$$(\sigma_{22})_{x_2=0,x_1=\ell+s} = \frac{N}{\sqrt{s}} + g(\ell) + O(s^{1/2}), \tag{6.6}$$

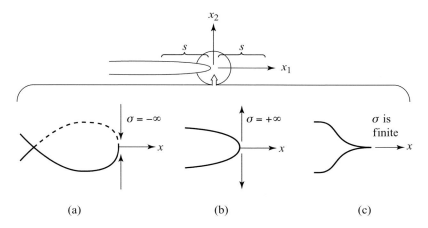

Figure 6.2 The formal solution to the elasticity problem under discussion is, generally speaking, singular near the cut tips; see the text.

where

$$N = \frac{\sqrt{2\ell}}{\pi} \int_0^\ell \frac{g(\xi)d\xi}{\sqrt{\ell^2 - \xi^2}} \, . \tag{6.7}$$

Furthermore, according to the first Kolosov–Muskhelishvili formula (6.2), on the cut we have

$$\frac{E}{1+\nu}(w_1 + iw_2) = (3 - 4\nu)\phi(x_1) - \overline{\phi(x_1)} \, ,$$

so that

$$\frac{E}{1+\nu}(\partial_1 w_1 + i\partial_1 w_2) = (3 - 4\nu)\Phi(x_1) - \overline{\Phi(x_1)}$$

and

$$D_{12} = \partial_1 w_2 = \mp \frac{4(1-\nu^2)}{E} \frac{N}{\sqrt{s}} + O(\sqrt{s}) \, . \tag{6.8}$$

Here the negative and positive signs correspond respectively to the upper and lower sides of the cut, so that

$$w_2 = \pm \frac{2(1-\nu^2)}{E} N\sqrt{s} + O(s^{3/2}) \, . \tag{6.9}$$

We come now to the important conclusion that, depending on the sign of N – this quantity is known in the literature as the *stress intensity factor* – the linear theory of elasticity offers the following possibilities (Figure 6.2).

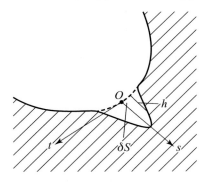

Figure 6.3 Variation of a crack contour formed at an area δS of the original surface.

(1) If N is negative then the theory of elasticity gives an obviously inappropriate result: the opposite sides of the cut penetrate inside each other (which is not formally prohibited by the problem formulation; the stress is prescribed on the cut and no restriction is posed on the displacement). At the tip the stress is negative and infinite. Obviously this solution cannot be an intermediate asymptotics of the deformation process.

(2) If N is positive, the stress at the cut tip is positively infinite. The obtained solution cannot be an adequate model of a crack because the material cannot sustain an infinite tensile stress.

(3) There remains the exceptional case $N = 0$. In this case (Figure 6.2c) the stress at the crack tip is finite, and a normal section of the cut has a characteristic cusp form. Moreover, the stress distribution at the crack surface and its extension is continuous in this case, contrary to the previous cases.

The *mobile-equilibrium* cracks in an ideally elastic body are distinguished from cuts and holes because generally speaking they do not remain intact if the loads are varying but can extend (or shorten, if the cracks can be healed). Therefore the position of the crack contour also should be obtained from the condition of the stationarity of the elastic energy.

Let us take an elastic body with a crack, for simplicity's sake a symmetric and plane crack, i.e. occupying a part of a plane under loading that is symmetric with respect to this plane. We will consider a virtual state of the body, under the same loads, which differs from the original state only by a small extension of the crack (Figure 6.3) near a given point O of the crack contour, i.e. by the formation of a new free surface at a small area δS. We emphasize that all loads remain invariant under such a crack extension, both with respect to their magnitude and their position. Up to small quantities, the relative normal displacement of opposite points on the crack sides of the new crack surface is, according to (6.9),

$$2w_2 \approx \frac{4(1 - v^2)}{E} N \sqrt{h - s},\qquad(6.10)$$

where h is the "height" of the additional crack area (see Figure 6.3). According to (6.6), the normal stress at a distance s from the crack contour is equal to $N/\sqrt{s}$ + small quantities of higher order. The energy δW released by the formation of the new crack area is equal to the work needed to close the extended crack. To close the crack it is sufficient to apply at its opposite sides the stress $N/\sqrt{s}$ which disappeared when the new crack surface opened. Therefore, up to small quantities of higher order we obtain

$$\delta W = \int_{\delta S} \sigma_{22} w_2 \, ds \, dt = \frac{4(1 - v^2)N^2}{E} \int_a^b dt \int_0^h \sqrt{\frac{h - s}{s}} \, ds$$

$$= \frac{2(1 - v^2)N^2\pi}{E} \int_a^b h \, dt = \frac{2(1 - v^2)\pi\delta S}{E} N^2.\qquad(6.11)$$

This very important formula (6.11) was obtained by G. R. Irwin (1957).

For a mobile-equilibrium crack in an elastic body the first variation of the elastic energy δW should vanish. From this a fundamental condition follows, valid for mobile-equilibruim cracks in an elastic body near their contour: *at the crack contour the stress is finite, the stress field is finite and continuous and the opposite sides of the crack close smoothly, forming a cusp-shaped normal section.* This condition, or postulate, was proposed by S. A. Christianovich (see Zheltov and Christianovich, 1955), and was named after him.[3]

It is precisely this condition comprising finiteness, continuity of the stress field and smooth closing of the crack surface at its contour that makes the problem of the elastic equilibrium of a body with cracks substantially nonlinear, in spite of the linearity of the basic equations. Indeed, the new part of the boundary of the body, involving the cracks, is not prescribed beforehand. According to this condition the cracks adjust to the acting load in such a way that the singularity of the elastic field at the crack contours disappears. More precisely, the problem of elastic equilibrium for a body with cracks can be formulated as follows (Barenblatt, 1956).

Suppose that an elastic body is given as well as the initial form of the cracks in it and also the loading process: the loads acting on the body are prescribed as functions of a parameter P. In the absence of pre-loading the initial state corresponds to P being equal to zero. The problem is to *find an elastic field and the shape of the crack surfaces for $P > 0$ that satisfy the equilibrium equations, the classic*

---

[3] In fact, there is a deep analogy between the Christianovich postulate and the Kutta–Chaplyguine–Joukovsky postulate (see Chapter 4).

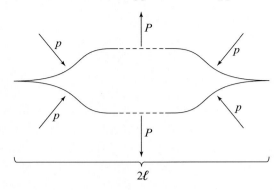

Figure 6.4 A crack, in a body under pressure, supported by two opposite concentrated forces.

boundary conditions and the conditions of stress finiteness and continuity and smooth closing of the crack contours.

Figure 6.4 shows an example: an isolated mobile-equilibrium crack, in an infinite body under pressure p, supported by two equal and opposite concentrated forces P at the centers of the crack sides. It is easy to perform a corresponding experiment: take a thick book, place it horizontally, and insert between the sheets a thin object, e.g. a nail. Pressure is created with weight of the book, and a crack-like cave with smoothly closing edges will appear. Putting some additional load on the book will shorten the crack.

Clearly, when the crack is closed the normal stress at its location under the applied loads will be equal to p, so (see Section 5.4.2)

$$G(x_1) - g(x_1) = P\delta(x_1) - p, \qquad (6.12)$$

where $\delta(x_1)$ is the Dirac δ-function. According to (6.7) we obtain

$$N = \frac{\sqrt{2\ell}}{\pi} \int_0^\ell \frac{[G(\xi) - g(\xi)]\, d\xi}{\sqrt{\ell^2 - \xi^2}} = \left(\frac{P}{\pi\ell} - p\right)\sqrt{\frac{\ell}{2}}, \qquad (6.13)$$

so the finiteness and continuity of stress at the crack contour and the smooth closure of the crack sides, i.e. the vanishing of N, is obtained when

$$\ell = \frac{P}{\pi p}. \qquad (6.14)$$

Relation (6.14), up to a constant factor, can be easily obtained using dimensional analysis. Indeed, the length of the crack ℓ depends on P and p only: the Christianovich condition does not add new arguments. So $[\ell] = L$, $[P] = F/L$ (the force distributed over the length) and $[p] = F/L^2$. The only quantity having the

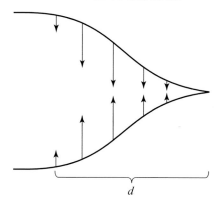

Figure 6.5 Cohesion forces.

dimension of length which can be obtained from the arguments P and p is P/p, therefore $\ell = \text{const} \times P/p$ and, according to (6.14), the constant is equal to $1/\pi$.

6.3 Cohesion crack model

Relation (6.14) is very natural indeed but it reveals a seeming paradox: at zero pressure, $p = 0$, and arbitrary small fracturing force P the length of mobile-equilibrium cracks appears to be infinite, so the body will be fractured by any small load if it contains even a small crack! This contradiction to everyday experience is due to the following fact: *we have not taken into account all forces acting on the body.* Indeed, the requirement for smooth closing of the cracks at their contours implies that there must be cohesion forces acting on the crack surface (Figure 6.5). These cohesion forces can be of various types.

In the ideal case of a purely brittle fracture, when plastic deformations are absent within the body, these forces are due to intermolecular cohesion. They are always present at the part of the fracture surface near the contour because the opposite sides of the cracks close smoothly. The intensity of the cohesive forces is equal to zero when the distance between the sides is the normal interatomic distance. They grow quickly with increasing distance, reaching very high values (of the order of the Young modulus), but after reaching a maximum they quickly decay.

Moreover, there exists a much wider class of *quasi-brittle* fracture phenomena, illustrated in Figure 6.6 by a photograph of a crack in polymethylmetacrilate (PMMA, a plastic widely used now in technology). In this case plastic deformation within the body exists, but it is concentrated near the crack surface and especially in the small "head" of the crack near its contour. Therefore the boundary between the elastic and plastic regions can be considered as effectively the crack surface, and

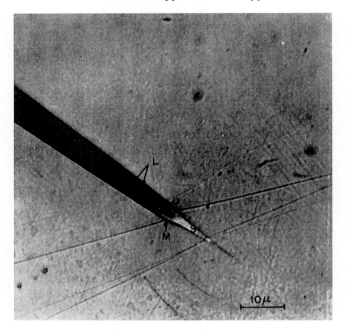

Figure 6.6 The crack tip in polymethylmetacrilate. Near the crack tip a white smoothly closing region is seen. From Van den Booghart (1966).

the forces that act on the elastic part from the plastic head of the crack can be considered as cohesion forces.

The possibility of using Griffith's approach (which was developed originally for ideally brittle fracture) to analyze quasi-brittle fracture was proposed by Irwin (1948) and Orowan (1950). This step was very important because purely brittle fracture is rather rare whereas quasi-brittle fracture often occurs in structural materials.

The quantitative laws governing the distribution of cohesion forces over the crack surface are even now practically unknown. However, the characteristic properties of cohesion forces, then large maximum intensity and fast decay with distance from the crack contour, allow us to introduce two hypotheses of a phenomenological nature and thus to construct an effective model, which has received in the literature the name *cohesion crack model* (Barenblatt, 1959); see also Barenblatt (1962, 1964), Goodier (1968) and Broberg (1999). We shall now discuss these two hypotheses.

First hypothesis (the hypothesis of smallness). The width d of the part of the crack surface (Figure 6.5) *over which the cohesion forces act is small in comparison with the crack size.*

In fact this hypothesis picks out a certain class of problems concerning the strength of structures with cracks which allow a simplified intermediate-asymptotic approach, presented below. In particular, this approach cannot be recommended for problems involving the generation of cracks. It is significant, however, that for a wide class of structural materials and structures, the dangerous cracks met with in practice do allow the use of the hypothesis of smallness.

Furthermore, under normal conditions the cohesion forces do not recover after unloading. At the beginning of the loading process for a body with cracks, the cohesion forces near the crack contours grow but the crack contours remain immobile, i.e. the cracks do not extend until the cohesion force reaches a maximum. When, at a certain location in the crack contour, the maximum value of the cohesion force is reached, the crack achieves *mobile equilibrium* and at this place begins to extend.

Second hypothesis (the hypothesis of autonomy). The heads of all mobile-equilibrium cracks, i.e. the form of the normal sections of the crack surface near its contour, and hence the distribution of cohesion forces over the crack surface, are identical for a given material and given external conditions.

According to the hypothesis of autonomy the crack head in the material plays the role of a "zipper", which moves under increasing load but in the mobile-equilibrium state always creates the same distribution of forces, stresses and displacement in its vicinity.

Obviously (cf. formula (6.7)), the stress intensity factor N is a linear functional of the applied loads, so it can be represented in the form

$$N = N_0 + N_c \,,$$

where N_0 is the stress intensity factor calculated without taking into account cohesion forces and N_c is the same quantity calculated for the cohesion forces only. To calculate N_c we use the hypothesis formulated above. According to the autonomy hypothesis we can calculate N_c for every special case and it will be the same for all cracks in the given material, under given external conditions. Therefore, we will calculate it for an isolated symmetric crack in an infinite body under a plane strain, using formula (6.7). For this special case, $g(x) = 0$ at $0 \le x \le \ell - d$ and $g(x) = -G(x)$ at $\ell - d \le x \le \ell$. Here $G(x)$ is the distribution of the cohesion forces over the normal section and d is, as before, the width of the area where the cohesion forces are acting. We have, using the hypothesis of smallness ($d \ll \ell$),

$$N_c = -\frac{\sqrt{2\ell}}{\pi} \int_{\ell-d}^{\ell} \frac{G(x)\,dx}{\sqrt{\ell^2 - x^2}} \approx -\frac{\sqrt{2\ell}}{\pi} \int_{\ell-d}^{\ell} \frac{G(x)\,dx}{\sqrt{2\ell x s}\,\sqrt{\ell - x}} = -\frac{1}{\pi} \int_0^d \frac{G(s)\,ds}{\sqrt{s}} ,$$

$$\tag{6.15}$$

where $s = \ell - x$. According to the autonomy hypothesis the integral on the right-hand side of (6.15) is a material constant, called the *fracture toughness* or *cohesion modulus*,

$$K = \int_0^d \frac{G(s)\,ds}{\sqrt{s}}. \tag{6.16}$$

This quantity, introduced by the present author (Barenblatt, 1959), characterizes the resistance of the material to crack extension and is a characteristic of the material strength. It should be distinguished from the strength characteristic K_{Ic}, introduced by Irwin (1960), which determines the beginning of catastrophic crack extension. Catastrophic crack extension requires the instability of the mobile-equilibrium state (for details, see below). At the beginning of crack extension from an unstable state an autonomous crack head has not been formed as yet, so a scatter of the data when this quantity is obtained experimentally is unavoidable.

According to the condition of the finiteness of stresses and the smooth closing of cracks, $N = N_0 + N_c = 0$, i.e. $N_0 = -N_c$. Thus we conclude that, at points of the contour where the mobile-equilibrium state is reached, the following condition must hold:

$$N_0 = \frac{K}{\pi}. \tag{6.17}$$

At points of the crack contour where the mobile equilibrium state is not reached the cohesion forces have not attained their maximum, so that $N_0 \leq K/\pi$. However, at these points the crack does not extend and thus the position of the crack contour is known beforehand.

We come now to a modification of the previous nonlinear elasticity problem statement for a body with cracks. A system of initial cracks in a body is given and also a loading process, i.e. a system of loads growing continuously from zero with increasing loading parameter P. The stresses, strains and displacements, as well as other elastic field characteristics, and *the shape of the cracks* should satisfy the elastic equilibrium equations, classical boundary conditions and the condition $N_0 \leq K/\pi$ at the crack contours. Note that if the cracks extend over curved surfaces then *owing to the autonomy hypothesis the direction of their extension at each point of the contour can be obtained from the condition of local symmetry of the elastic field.*

Now we make an important comment concerning the energy approach to the problem of brittle and quasi-brittle fracture. In our previous analysis we used the force approach: the cohesion forces acting at the edge regions of cracks were introduced explicitly. Griffith (1920, 1924) and, following him, Irwin (1957) used a different approach, the energy approach, introducing directly an integral characteristic of the cohesion forces and the surface tension γ, i.e. the energy required for formation of the unit area of the crack surface. On the one hand, according to Irwin's formula (6.17) the release of elastic energy δW, calculated without taking into account the cohesion forces, is equal to

$$\delta W = \frac{2(1 - v^2)\pi \delta S}{E} N_0^2 .$$

On the other hand,

$$\delta W = 2\gamma \delta S .$$

The factor 2 is due to the fact that two fracture surfaces are formed. Relation (6.17) and the last two relations make it possible to obtain an expression for the cohesion modulus K in terms of the surface tension γ and the elastic constants E and v:

$$K^2 = \frac{\pi E \gamma}{1 - v^2} .$$

Subsequent developments in fracture models (Leonov and Panasyuk, 1959; Dugdale, 1960; Panasyuk, 1968; and many later publications) have shown that the force approach has certain advantages, especially for the modeling of quasi-brittle fracture. It allows one to take into account explicitly the properties of "bonds" in the edge regions of cracks. In an important paper (Willis, 1967), a comparison of the force and energy approaches was performed.

6.4 What is fracture from the mathematical viewpoint?

Let us consider the special case when the symmetry of the body and of the applied loads makes possible the propagation of rectilinear cracks under plane strain conditions. In the case of an isolated crack in an infinite body the expression for N_0 is given by formula (6.7). If all the loads are proportional to the loading parameter P then the normal stress $g(x)$ at the location of the crack under consideration should also be proportional to P : $g(x) = Pn(x)$, where the function $n(x)$ does not depend on P. Therefore, if the mobile-equilibrium state is achieved at the crack tips, we obtain from (6.17) a *finite equation* for the crack size ℓ:

$$\frac{P}{K} = \left(\sqrt{2\ell} \int_0^\ell \frac{n(\xi)d\xi}{\sqrt{\ell^2 - \xi^2}} \right)^{-1} = f(\ell), \qquad (6.18)$$

where $f(\ell)$ can be considered to be a known function of ℓ. When the mobile-equilibrium state at the crack tip is not reached, P/K is less than $f(\ell)$, but the crack length ℓ_0 remains the same.

An instructive example will show us the nature of fracture from the mathematical viewpoint.

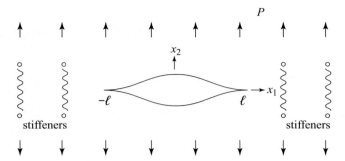

Figure 6.7 Symmetric crack in a body with two pairs of stiffeners under uniform tension.

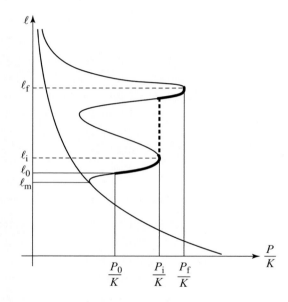

Figure 6.8 Crack extension in a body with two pairs of stiffeners under a growing load. After reaching a critical length size, ℓ_f, and critical load, P_f, the solution to the nonlinear problem ceases to exist.

Let us consider an infinite body with a single symmetric crack and two pairs of stiffeners (Figure 6.7). If the body is loaded at infinity by a homogeneous tensile stress P, the graph of the function $f(\ell)$ has a characteristic two-hump form (Figure 6.8).

The shape of the curve presented in Figure 6.8 can be easily obtained using the examples in Sections 5.4.1 and 5.4.2. Indeed, $n(\xi) = n_1(\xi) + n_2(\xi) + n_3(\xi)$. The load distribution $n_1(\xi)$ corresponds to a homogeneous load, so that $n_1(\xi) = 1$. In the absence of stiffeners, i.e. when $n_2(\xi) = n_3(\xi) = 0$,

$$f(\ell) = \frac{\sqrt{2}}{\pi\sqrt{\ell}} . \tag{6.19}$$

The functionals $n_2(\xi)$ and $n_3(\xi)$ can be obtained in the following way. For each pair of stiffeners the load is taken as corresponding to four concentrated forces X_2^j proportional to P with different values of the coefficients of proportionality (stiffnesses) and directed along the x_2 axis, so that the Kolosov–Muskhelishvili potentials for them are equal, according to formula (5.68):

$$\Phi_j(z) = \frac{iX_2^j}{2\pi(1+\kappa)} \left(\frac{1}{z-\ell_j-im_j} - \frac{1}{z-\ell_j+im_j} + \frac{1}{z+\ell_j-im_j} - \frac{1}{z+\ell_j+im_j} \right),$$

for $j = 1, 2$, where ℓ_j is the horizontal distance of the jth pair of stiffeners from the symmetry line formed by the x_2 axis and m_j is the vertical distance of the jth pair of stiffeners from the symmetry line formed by the x_1 axis. Furthermore, $\Psi_j(z) = -\kappa\Phi_j(z)$. Taking various combinations of X_2^j and ℓ_j, m_j and using formulae (5.61) we can easily obtain the function $f(\ell)$ for this problem.

Therefore the initial crack size ℓ_0 remains invariant until the load reaches the value P_0 when the crack becomes a mobile-equilibrium one. In the case under consideration, for which $\ell_m < \ell_0 < \ell_i$ (Figure 6.8), the crack begins to extend slowly with growing load after the load reaches the value P_0. *This beginning of growth is not yet fracture, because the body with the crack continues to support the growing load.* When the load reaches the value P_i, however, a bifurcation of the mobile equilibrium takes place: the crack extends by a jump, breaking through the first pair of stiffeners. The corresponding point in the $(P/K, \ell)$ plane jumps to another branch of the curve $f(\ell)$. *The jump is also not yet a fracture: after it the crack continues to grow slowly but the cracked body continues to support the growing load* until the load P_f is reached, with corresponding crack length ℓ_f, at which the crack breaks the second stiffener. A body with a crack of size ℓ_f reached in the loading process is no longer able to support growing loads; the cohesion forces at the crack edges are insufficient to support such a large crack. Mathematically, this means that a solution to the equation (6.18) at $P > P_f$ satisfying the condition $\ell > \ell_f$ does not exist.

In the absence of stiffeners there will be no humps on the curve $P/K = \phi(\ell)$. Therefore (Figure 6.9) the crack remains invariant for $P < \sqrt{2}K/(\pi\sqrt{\ell_0})$. When P reaches $P_f = \sqrt{2}K/(\pi\sqrt{\ell_0})$, a catastrophic fracture occurs: the solution to equation (6.18) ceases to exist. In this case a stage of stable crack growth preceding fracture does not exist. An autonomous crack head has not yet formed. Therefore, using the formula $K = P_f\pi\sqrt{\ell_0}/\sqrt{2}$, as is done when the parameter K_{Ic} is obtained from

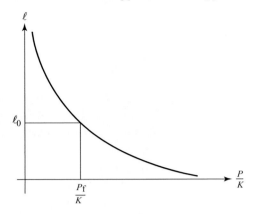

Figure 6.9 The behavior of a body with a crack without stiffeners. When the load P is less than $P_{\mathrm{f}} = \sqrt{2}K/(\pi \sqrt{\ell_0})$ the crack remains intact. For $P = P_{\mathrm{f}}$ a catastrophic fracture occurs.

experiment, can lead, as mentioned above, to a large scatter of the data; this can be avoided by using an experimental procedure with stable crack growing.

As an example of such a procedure, we mention the case of a large body containing a crack supported by opposite concentrated forces P (distributed over the straight lines $x_1 = 0$, $x_2 = \pm 0$). In this case $n(\xi) = \delta(x_1)$, where $\delta(x)$ is the Dirac δ-function, so that $f(\ell) = \sqrt{2\ell}$ with

$$\ell = \frac{P^2}{2K^2} . \tag{6.20}$$

Measuring P and ℓ simultaneously during this stable process of crack growth, the value of the fracture toughness, or cohesion modulus, can be obtained: $K = P/\sqrt{2\ell}$. Formula (6.20) can be easily obtained, up to a constant factor, by dimensional analysis. As formula (6.16) shows, the dimension of the cohesion modulus, $[K]$, is equal to the dimension of stress times the dimension of the square root of length: $[K] = FL^{-3/2}$. The crack length depends on P and K only (remember that in this case the extra pressure p is absent). Therefore the only quantity having the dimension of length that can be obtained from P and K is P^2/K^2; the constant remains undetermined.

Another example of stable crack propagation is given by the remarkable experiments of Roesler (1956) and Benbow (1960). A punch with a small flat point was pressed slowly into the face of a cubic block of transparent brittle material (fused silica; see Figure 6.10). A perfect conical crack was formed at an intermediate stage, with the base diameter of the conical crack larger than the diameter of

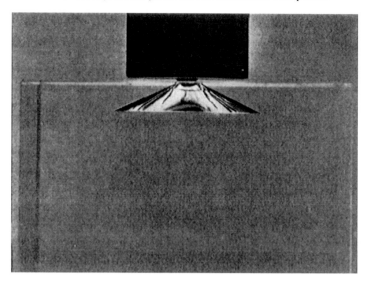

Figure 6.10 When a punch is pressed into a block of fused silica, a conical crack is formed (Benbow, 1960).

the flat point of the punch but smaller than the size of the block. As the load was increased, the crack increased in size.

In this problem neither an analytical nor a numerical solution was obtained. The only way to obtain a quantitative description of this phenomenon is with dimensional analysis. Here again it is natural to assume that at the intermediate stage the diameter D of the base of the conical crack depends on the load P and the properties of the material, its cohesion modulus K (the crack propagation is stable) and Poisson ratio v. The Young modulus E does not enter the set of governing parameters because the loads, not the displacements, are prescribed.

For the intermediate stage under consideration it is natural to assume that neither the diameter of the punch nor the block size enters the list of governing parameters, so that $D = f(P, K, v)$. The dimensions of these quantities are $[D] = L$, $[P] = F$, $[K] = FL^{-3/2}$, $[v] = 1$. Dimensional analysis immediately yields

$$D = f(v)\left(\frac{P}{K}\right)^{2/3} . \tag{6.21}$$

An analysis of the experimental data (Figure 6.11) confirmed this relation. This experiment was also used to determine the cohesion modulus.

A numerical investigation of this problem, which has not yet been performed, might be instructive. A thin ring-formed cylindrical notch of small height under the edge of a patch should be prescribed, and the load (in fact, the uniform displacement) should gradually be increased. When the stress intensity factor N at

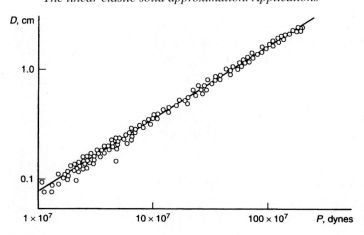

Figure 6.11 The experimental data for the propagation of a conical crack in a block of fused silica confirm the scaling law $D = \text{const} \times (P/K)^{2/3}$ (Benbow, 1960).

the edge of the notch reaches the critical value of K/π, the notch will become a crack and, when the load is gradually increased further, the direction of the crack should be selected using the condition of local symmetry, which follows from the hypothesis of autonomy. The data from the experiment of Roesler and Benbow (Figure 6.10) suggest that soon afterwards the intermediate self-similar stage considered above would be reached. The computation should continue, and when the diameter of the base of the crack, D, increases sufficiently (the computation can be performed in a finite region as always), the deviation of the crack from a conical shape should begin.

The most interesting point is as follows: if the computation continues, at a certain step the numerical procedure is expected to fail. *This would constitute a numerical proof of the non-existence of the solution of the problem of slow crack extension for larger loads: further crack extension will be fast, and the static equilibrium equations used in the first stage will be inadequate.*

Returning to the problem considered above of a crack with two pairs of stiffeners, we recall that the same situation happened (see Figure 6.8) when the break of the crack through the first pair of stiffeners occurred. However, in this case, owing to symmetry, it is suggested that the fast jump of the crack could terminate at the point where the second pair of stiffeners creates the same stress intensity factor. Therefore there would be no need to consider the dynamic stage of crack propagation. This is not the case for the conical crack.

Moreover, in a real situation, where the structure contains many cracks, the static stage of development of the system of cracks terminates rather quickly. This does not necessarily mean the fracture of the structure; there may only be a transition to the dynamic stage, when the development of the system of cracks is described by the dynamic equations of elasticity. However, in this case the problem is not properly formulated as yet because the conditions at the crack tip do not allow us, generally speaking, to make any kind of autonomy hypothesis. The obvious reason for this is that the velocity of propagation of elastic waves is high and, a priori, at the dynamic stage the influence of the boundaries cannot be excluded.

We mention in conclusion that, as previous considerations have shown, the local criteria for fracture of the Huber–von Mises–Hencky type mentioned before are inadequate. It has been shown by the examples just considered that such criteria do not work for bodies with cracks: the stresses at the tips of the cracks, calculated without taking into account the cohesion forces, are always infinite. Moreover, the stress concentration at the tips of defects having a small although finite radius of curvature (which are always present in structures) can be arbitrarily large, although these defects are not dangerous.

We emphasize that fracture is not related to a local fact, that of reaching some critical condition at a certain point in a structure. Just the opposite: fracture (or at least the beginning of the dynamic stage) is related to the global fact of the loss of existence of the solutions to the nonlinear problem of the elastic equilibrium of a body with cracks, formulated above.

We note two points in conclusion. Fracture belongs to the class of problems where the basic interest is not in the solution itself but in a seemingly pure mathematical question, the range of parameter values where a solution does exist. Indeed, the stress, strain, displacement etc. fields are of much less interest for a consumer than the limiting value P_f of the loading parameter after which fracture takes place and the solution ceases to exist. Furthermore, we have seen that the limiting load at $\ell_0 < \ell_f$ happens to be independent of the initial crack size ℓ_0. This independence is a general fact if, between the beginning of the crack extension and fracture, there exists a stage of stable crack extension when the mobile-equilibrium crack extends slowly and continuously with increasing load.

6.5 Time effects; lifetime of a structure; fatigue

The considerations presented above do not take into account the time factor: according to the accepted model the limiting loads remain as such for all times. As everyday experience shows, this is not so: fracture often occurs not immediately but some time after application of a load. The reason for this discrepancy is that we did not take into account the deterioration with time of the bonds that create the cohesion forces near the crack contour. If this deterioration is taken into account, the fracture toughness ceases to be a material constant.

In practice the basic case is that of quasi-steady crack propagation when, during crack extension over a distance equal to the width of the cohesive zone d, the variation in the stress intensity factor is negligible. In this case the fracture toughness becomes not a constant but a universal functional of the velocity of the crack extension u. Usually, when the process of crack extension is slow it may be assumed that the cohesion modulus is a universal function $K = K(u)$ (for a given material and given external conditions) of the crack tip velocity u.

For a wide class of materials the relation $K(0) = 0$ can be assumed. The dependence of the cohesion modulus on crack extension rate completely changes the very statement of the fracture problem (Barenblatt, Entov and Salganik, 1966). Indeed, relation (6.18) ceases to be a finite equation determining the length of a mobile-equilibrium crack. In the simplest case under consideration, that of an isolated rectilinear crack in an infinite body, this relation becomes an ordinary differential equation (note that $u = d\ell/dt$),

$$\frac{P}{K(d\ell/dt)} = f(\ell) \,, \tag{6.22}$$

which determines the crack's half-size $\ell(t)$ as a function of time for a given initial value ℓ_0 and given loading process $P(t)$.

For the special case when $P(t)$ is identically equal to a constant, the function $K(u)$ grows monotonically and $K(\infty) < \infty$, the solution to the problem can be conveniently illustrated graphically. For a crack in an infinite body with two pairs of stiffeners, the solution is shown in Figure 6.12.

Starting from the value ℓ_0 (ℓ_0 is the initial crack size) the curve $P/(Kd\ell/dt)$ follows smoothly the curve $f(\ell)$ (see (6.22)). The value of $d\ell/dt$ is determined at every point (see Figure 6.12) without any irregularities until ℓ reaches the value ℓ_1, where $f(\ell_1) = P/K(\infty)$. At this point the continuous growth of the crack size stops and a jump follows. It can be interpreted as the breaking of the crack through the first pair of stiffeners. After that the crack tip speed starts to diminish until the second maximum of $f(\ell)$ is reached, at $\ell = \ell_{m2}$. After reaching a minimum the crack tip speed starts to grow, and when ℓ reaches the value ℓ_f, where $f(\ell_f) = P/K(\infty)$, the solution ceases to exist.

Here again, neither the beginning of crack extension nor the start of catastrophic crack extension necessarily mean fracture. Fracture is as before related to the loss of existence of a solution to the problem of the elastic equilibrium of a body with cracks. The solution to equation (6.22) only exists for a finite time, generally speaking, even for an arbitrarily small load, so that every load ultimately produces fracture. The problem is thus not one of determining the fracture load but one of

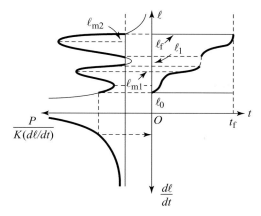

Figure 6.12 Crack extension in a body with two pairs of stiffeners under a constant load with time effects accounted for. There is no critical load: every load fractures. The problem is to determine the lifetime t_f after which the solution ceases to exist.

determining the *lifetime* of a given structure under a given load or, alternatively, of determining the limiting load which a given structure can carry during a prescribed lifetime. It is obvious that the lifetime and time of existence of a solution will always depend on the initial crack size.

It is noteworthy that for several materials the function $K(u)$ is non-monotonic and even non-single-valued. In such materials a loss of stability of uniform crack propagation occurs and the crack begins to propagate by jumps.

A very important phenomenon, also related to time effects, is that of *fatigue*. This phenomenon and its engineering applications are presented in a fundamental book by Suresh (1998). A standard fatigue tensile experiment is performed as follows (Figure 6.13). A specimen (a notched bar or plate) is loaded with a combination of a static tensile load and a pulsating tensile load of constant frequency and amplitude. At the tip of the notch a fatigue crack is formed, and its propagation, i.e. its length as a function of the number of cycles of the pulsating load, is recorded. For multi-cycle fatigue tests, when the number of cycles is of the order of many millions before failure, the following power law, known in the literature as *Paris' law*, was found to be established after a relatively short initial stage (see Paris and Erdogan, 1963):

$$\frac{d\ell}{dn} = A(\Delta N)^m .$$

(6.23)

Here n is the number of cycles, the value of $d\ell/dn$ taken to be the average over a cycle and $\Delta N = N_{\max} - N_{\min}$ is the amplitude of stress intensity factor. Under these experimental conditions we have $N = c\sigma \sqrt{\ell}$, where σ is the pulsating bulk

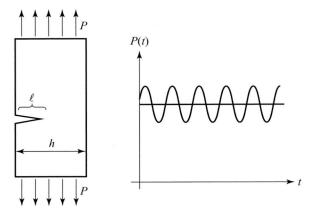

Figure 6.13 The fatigue test: principles of the scheme.

stress (the load divided by the undamaged specimen's cross-sectional area) and the constant c is a form factor, which can be evaluated using a theory-of-elasticity technique. The Paris law (6.23) has found multiple confirmation for different metals (see Figure 6.14) and is now considered to be one of the fundamental laws of strength engineering science. The exponent m has been found to vary in a wide range from slightly more than 2 to 10.

The Paris law is used as an important predictor of the lifetime of a structure, i.e. the number n_f of cycles before failure. Relation (6.23) can be rewritten, bearing in mind that $N = c\sigma \sqrt{\ell}$, in the form

$$\frac{d\ell}{dn} = Ac^m(\sigma_{\max} - \sigma_{\min})^m \ell^{m/2} . \tag{6.24}$$

By integration we obtain

$$\frac{1}{\ell_0^{m/2-1}} - \frac{1}{\ell^{m/2-1}} = G(m-2)n , \tag{6.25}$$

where $G = Ac^m(\sigma_{\max} - \sigma_{\min})^m/2$ and ℓ_0 is the initial crack length. In multi-cycle fatigue the number of cycles before failure is very high. Therefore in the evaluation of the lifetime it is possible to neglect the number of cycles corresponding to the preliminary stage when the Paris law (6.23) does not hold. So, because the intermediate stage, where the Paris law is valid, holds during the basic part of the fatigue test, an estimate for the lifetime, i.e. for the number n_f, can be obtained from (6.25), assuming that $\ell \gg \ell_0$ and neglecting the second term on the left-hand side of (6.25):

$$n_f = \frac{1}{(m-2)G\ell_0^{(m-2)/2}} . \tag{6.26}$$

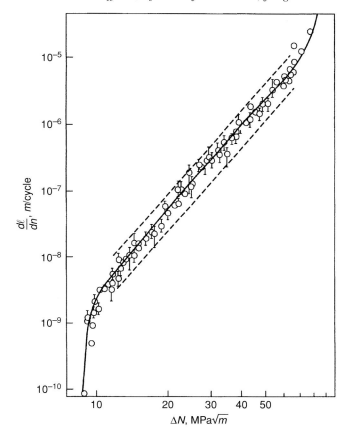

Figure 6.14 Experimental data for fatigue crack growth in a titanium alloy confirm Paris' law (6.23) in the major, intermediate, part of the crack velocity range (Botvina, 1989).

Clearly the lifetime n_f depends sharply upon the parameter m, the exponent entering the Paris law, and also upon the pre-power coefficient A entering the expression (6.23) of this law.

Therefore it is crucially important to find out whether these constants are universal material characteristics, i.e. first of all, whether they are the same for a given material under given conditions for specimens of various sizes h. Otherwise, using the results of standard fatigue tests performed on small specimens could be dangerous in practical structure design: the real lifetime of the structure could be overestimated. We will return to this important question later, in Chapter 9.

7

The Newtonian viscous fluid approximation. General comments and basic relations

7.1 The fundamental idealization. The Navier–Stokes equations

Consider a simple shear flow, the flow of fluid between parallel plates (Figure 2b in the Introduction); where the longitudinal size of the plates is much larger than the distance between the plates. The lower plate is at rest, the upper one is moving with velocity V (see Figure 7.1a). For an important class of fluids (water, air, glycerin at room temperature), experiment shows that the velocity distribution between the plates is linear. Such an experiment, performed originally by Newton, demonstrates that the fluids have the property of *viscosity*; owing to the thermal oscillations of the molecules of the fluid a transfer of momentum in the transverse direction occurs in the flow, and therefore tangential forces appear. Newton's law (see also the Introduction) determines the density of the transverse momentum flux, i.e. the shear stress τ:

$$\tau = \eta \, \frac{V}{h} \, . \tag{7.1}$$

In fact, an important message implicit in this law is that the quantity η, the dynamic viscosity coefficient, does not depend on the velocity V or the thickness of the fluid layer h; at fixed external conditions (temperature, pressure) it is a constant fluid property.

We note that for many fluids, especially polymeric melts and concentrated polymeric solutions, clay muds used in drilling etc., the coefficient η is no longer a constant fluid property. Physically this is explained by the fact that these fluids have a supramolecular microstructure that changes in the process of flow. However, it is important that there exists a sufficiently wide class of fluids and flows for which the *dynamic coefficient of viscosity* is constant in a wide range of values of the velocity gradient V/h. After being determined in one experiment this coefficient can be used to describe any motion of the fluid.

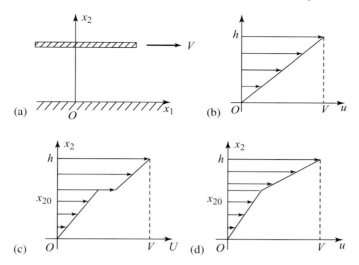

Figure 7.1 (a) Newton's experiment. (b) A continuous smooth linear velocity distribution. (c) A discontinuous velocity distribution. (d) A velocity distribution with a discontinuous derivative (a kink).

Passing to a general three-dimensional model based on Newton's law of fluid motion, we remind the reader that, as shown in Section 3.2, the velocity field $\mathbf{u}(\mathbf{x} + \mathbf{r})$ in a small vicinity of a given point $\mathbf{x}$ can be represented as a composition of the velocity of translational movement of the center of the vicinity $\mathbf{u}(\mathbf{x})$, the velocity of rotation around the center with angular speed $\boldsymbol{\omega} = \frac{1}{2}\operatorname{curl}\mathbf{u}$ (these two components give the velocity of the motion of the whole vicinity as a solid body), and a third velocity component. This last component, $\mathbf{D} \cdot \mathbf{r}$, determines the velocity of deformation. Here

$$\mathbf{D} = \tfrac{1}{2}(\operatorname{grad}\mathbf{u} + \operatorname{grad}\mathbf{u}^{\mathrm{T}}) \tag{7.2}$$

is a second-rank tensor, *the strain-rate tensor*. The tensor $\mathbf{D}$ is symmetric, therefore it has three real eigenvalues and three principal directions, determined by the orts (unit vectors) – eigenvectors of $\mathbf{D} - \mathbf{e}_1, \mathbf{e}_2, \mathbf{e}_3$. The small vectors $dx_i \mathbf{e}_i$ are extended or compressed during the time dt by an amount $\partial_i u_i dx_i dt$. However, the angles between these vectors remain at $90°$. Therefore the variation $dx_1 dx_2 dx_3$ of the volume V of a parallelepiped constructed on these vectors is (per unit time) equal to $\Delta V = V \partial_\alpha u_\alpha$, so that the rate of relative volume variation is

$$\frac{\Delta V}{V} = \partial_\alpha u_\alpha = \operatorname{trace}\mathbf{D} . \tag{7.3}$$

Furthermore, the components $D_{ij} = \frac{1}{2}(\partial_j u_i + \partial_i u_j)$ of the strain-rate tensor $\mathbf{D}$ in a Cartesian system oriented arbitrarily with respect to the basic one defined by $\mathbf{e}_1$, $\mathbf{e}_2$, $\mathbf{e}_3$ have the following clear interpretation (cf. the same argument in Section 5.1 for a solid body). For $i = j$, D_{ij} is the rate of extension (or contraction) of the infinitesimal vectors $dx_i \mathbf{e}_i$ along the axis x_i. For $i \neq j$, D_{ij} is one-half the skewing rate of what are initially right angles between the vectors $dx_i \mathbf{e}'_i$ and $dx_j \mathbf{e}'_j$. (The orts $\mathbf{e}'_i$ determine the arbitrary Cartesian system.)

Let us turn now to the stress tensor $\boldsymbol{\sigma}$. We introduce the following idealization, called the Newtonian viscous fluid approximation. Earlier, we separated the pressure term in the stress tensor $\boldsymbol{\sigma}$, assuming that it can be determined independently, writing

$$\boldsymbol{\sigma} = -p\mathbf{I} + \boldsymbol{\tau} . \qquad (7.4)$$

For a Newtonian viscous fluid the *constitutive equation*, relating the tensor $\boldsymbol{\tau}$ and the strain-rate tensor $\mathbf{D}$, has the form of a linear relation:

$$\boldsymbol{\tau} = 2\eta[\mathbf{D} - \tfrac{1}{3}(\mathrm{trace}\ \mathbf{D})\mathbf{I}] + \zeta(\mathrm{trace}\ \mathbf{D})\mathbf{I} . \qquad (7.5)$$

Here η, ζ are assumed to be universal constants for a given medium under given external conditions.

We emphasize that our arguments here repeat exactly the arguments in Section 5.1 for the case of an ideally elastic body: therefore we have omitted some details. In particular, the deviator $\mathbf{D} - \tfrac{1}{3}(\mathrm{trace}\ D)\mathbf{I}$ is taken instead of the tensor $\mathbf{D}$ itself in order to separate the contributions to the total stress of the shear strain rate and of the volume extension (or contraction) strain rate. The coefficients η and ζ are named accordingly the dynamic viscosity coefficient (having in mind the shear viscosity) and the volume viscosity coefficient.

Now we substitute the constitutive equation (7.5) into the equation for momentum balance (1.30):

$$\rho \frac{d\mathbf{u}}{dt} = \rho \mathbf{F} + \nabla \cdot \boldsymbol{\sigma} .$$

Thus we obtain

$$\rho[\partial_t \mathbf{u} + (\mathbf{u} \cdot \nabla) \cdot \mathbf{u}] = \rho \mathbf{F} - \nabla p + \eta \Delta \mathbf{u} + (\zeta + \tfrac{1}{3}\eta)\ \mathrm{grad}\ \mathrm{div}\ \mathbf{u} . \qquad (7.6)$$

(The transformations performed here are exactly the same as in the case of the linear elastic body, considered in Chapter 5, so again we have not repeated them.)

In the case of a viscous fluid of constant density ρ, the velocity field is solenoidal, $\mathrm{div}\ \mathbf{u} = \nabla \cdot \mathbf{u} = 0$, and equation (7.6) assumes the form

$$\partial_t \mathbf{u} + (\mathbf{u} \cdot \nabla) \cdot \mathbf{u} = \mathbf{F} - \frac{1}{\rho}\nabla p + \nu \Delta \mathbf{u} , \qquad (7.7)$$

where the quantity $\nu = \eta/\rho$ is called the *kinematic viscosity coefficient* (because its dimension $[\nu] = L^2 T^{-1}$ does not contain the dimension of mass, in contrast with the dynamic viscosity coefficient for which $[\eta] = ML^{-1}T^{-1}$).

The vector equation (7.7) is called the Navier–Stokes equation. It was obtained (in coordinate form) by the French scientists C. L. M. H. Navier (1822), A. L. Cauchy (1828) and S. D. Poisson (1829); their considerations were based on certain models of interaction of the molecules in matter. The equation was derived purely phenomenologically by the French scientist, A. J. C. Barré de Saint-Venant (Saint-Venant, 1843), a disciple of Navier and independently by the English applied mathematician G. G. Stokes (1845). All the scientists involved are well known in the history of applied mathematics and mechanics but traditionally only the names of the first- and last-mentioned appear in the name of this fundamental equation. Together with the equation of continuity,

$$\operatorname{div} \mathbf{u} = 0, \tag{7.8}$$

the Navier–Stokes equation (7.7) presents a closed system of equations for a Newtonian viscous incompressible fluid of constant density ρ.

Several comments can be made about the physical meaning of the assumptions forming the basis of the constitutive equation (7.5). The medium is assumed to be isotropic, so that the coefficients η and ζ are scalars; in general the invariance principle allows these quantities to be tensors. However, tensors of any order except zero (scalars) have principal directions, so that not all directions in the medium are equivalent. This contradicts the assumption of isotropy of the medium. Also, the medium is assumed to be homogeneous, so that the quantities η and ζ do not depend on the coordinates, i.e. they are constants of the material under given external conditions.

Furthermore, the constitutive equation (7.5) does not include higher velocity derivatives, i.e. strain-rate tensors of higher orders (tensors that are analogous to the strain-rate tensor $\mathbf{D}$ but which include higher velocity derivatives). If such tensors entered the linear constitutive equation then their scalar coefficients would have the dimension of the viscosity coefficients times the dimension of length. Therefore the medium would have a structural parameter, a characteristic length related to the internal microstructure of the material. So, in the absence of such terms we are neglecting the effects of the medium's internal microstructure. When considering the propagation of waves of very small wavelength, comparable with the characteristic length scale of the microstructure, the effects of higher-order terms could be substantial and the present approximation would be insufficient.

Solving the system (7.7), (7.8) is nowadays one of the greatest challenges in applied mathematics and fluid mechanics. Sometimes it is compared with climbing Mt Everest.[1]

---

[1] This mountain was named after Sir George Everest, who, by the way, was the great-uncle of the eminent British applied mathematician Sir Geoffrey Taylor, whose name the reader will meet very often in this book.

7.2 Angular momentum conservation law

By definition the strain-rate tensor **D** is symmetric. The constitutive equation (7.5) therefore ensures the symmetry of the stress tensor $\boldsymbol{\sigma}$. It is instructive to clarify what it means physically.

Let us construct the equation for angular momentum balance. It is more convenient and also clearer to do this in coordinates. We introduce the *alternating* Levi-Civita third-order tensor $\boldsymbol{\varepsilon}$ having components ε_{ijk} equal to unity if i, j, k are all different and follow a direct order, i.e. 1, 2, 3 or 2, 3, 1 or 3, 1, 2. If they follow an inverse order, e.g. 1, 3, 2, then $\varepsilon_{ijk} = -1$; if two of the indices i, j, k coincide then $\varepsilon_{ijk} = 0$.

Using this alternating tensor we can present the ith component of the angular momentum of an element $d\omega$ of the body in the form

$$\rho\varepsilon_{i\alpha\beta}x_\alpha u_\beta \, d\omega , \tag{7.9}$$

where summation from 1 to 3 over the indices α and β is, as usual, assumed. The rate of variation of the ith component of the angular momentum of the body when its instantaneous configuration is Ω (as before Ω is fixed in the observer's system of reference and bounded by a surface $\partial\Omega$) is given by

$$\partial_t \int_\Omega \rho\varepsilon_{i\alpha\beta}x_\alpha u_\beta \, d\tau . \tag{7.10}$$

In the absence of external forces and couples (7.10) consists of three parts. The first is the ith component of the flux of the angular momentum through the body's boundary surface $\partial\Omega$, leaving the body together with the medium flow:

$$- \int_\Omega \rho\varepsilon_{i\alpha\beta}x_\alpha u_\beta u_\gamma n_\gamma \, d\Sigma . \tag{7.11}$$

The second part is the angular momentum of internal forces, the tractions acting on the surface $\partial\Omega$:

$$- \int_{\partial\Omega} \varepsilon_{i\alpha\beta}x_\alpha \sigma_{\beta\gamma} n_\gamma \, d\Sigma . \tag{7.12}$$

The action of the ambient medium, and this is important, does not reduce to tractions. Also, the "internal couples" acting on a body due to the external medium should be taken into account in the equation for angular momentum balance; following the invariance principle this requires the introduction of a second-order tensor **M** with components M_{ij} (cf. the introduction of the stress tensor in the equation for momentum balance in Chapter 1). The ith component of the angular momentum vector of the couples acting on a unit area having normal **n** is $M_{i\gamma}n_\gamma$. Therefore the third part contributing to the variation (7.10) of the angular momentum is

$$-\int_{\partial\Omega} M_{i\gamma} n_\gamma \, d\Sigma \, , \tag{7.13}$$

so that the equation for angular momentum balance takes the form

$$\partial_t \int_{\Omega} \rho \varepsilon_{i\alpha\beta} x_\alpha u_\beta \, d\tau = -\int_{\partial\Omega} (\rho \varepsilon_{i\alpha\beta} x_\alpha u_\beta u_\gamma - \varepsilon_{i\alpha\beta} x_\alpha \sigma_{\beta\gamma} - M_{i\gamma}) n_\gamma \, d\Sigma \, . \tag{7.14}$$

From equation (7.14), using a Gauss transformation of the surface integral to a volume integral and the arbitrariness of the body configuration Ω, exactly as we did in deriving the equation for the momentum balance, we obtain a differential equation for the angular momentum balance:

$$\partial_t \rho \, \varepsilon_{i\alpha\beta} x_\alpha u_\beta + \partial_\gamma [\varepsilon_{i\alpha\beta} x_\alpha (\rho u_\beta u_\gamma - \sigma_{\beta\gamma}) - M_{i\gamma}] = 0 \, . \tag{7.15}$$

For the present case the *momentum balance equation* can be written in coordinates as

$$\partial_t \rho \, u_\beta + \partial_\gamma (\rho u_\beta u_\gamma - \sigma_{\beta\gamma}) = 0 \, . \tag{7.16}$$

Further, we multiply (7.16) by $\varepsilon_{i\alpha\beta} x_\alpha$ and subtract the equation obtained from (7.15). Note that $\partial_\gamma x_\alpha = \delta_{\gamma\alpha}$ (the Kronecker delta) and that the expression $\rho u_\beta u_\gamma \varepsilon_{i\alpha\beta} \delta_{\alpha\gamma}$, as it is easy to check, is equal to zero. We obtain

$$\varepsilon_{i\alpha\beta} \sigma_{\beta\gamma} = \partial_\gamma M_{i\gamma} \, . \tag{7.17}$$

Also $\varepsilon_{i\alpha\beta} \sigma_{\beta\gamma} = \sigma_{pq} - \sigma_{qp}$, where p and q are order indices from the sequence 1, 2, 3 and do not equal i. We have obtained that the stress tensor $\boldsymbol{\sigma}$ is symmetric only if the tensor of the internal couples $\mathbf{M}$ is solenoidal, i.e. $\nabla \mathbf{M} = 0$, in particular, if $\mathbf{M}$ is equal to zero. The effects of the internal couples, leading to asymmetry of the stress tensor, are seen in particular in suspensions of oblong particles in dielectric fluids in strong magnetic fields and also in blood. In such cases taking into account the internal couples is necessary.

7.3 Boundary value and initial value problems for the Newtonian viscous incompressible fluid approximation. Smoothness of the solutions

The system of the Navier–Stokes and continuity equations requires, for one to obtain a unique solution, a prescription of the initial condition for the velocity (but not for the pressure because the time derivative of pressure does not enter the system of equations),

$$\mathbf{u}(\mathbf{x}, t_0) = \mathbf{u}_0(\mathbf{x}) \, , \tag{7.18}$$

and also some boundary conditions. We emphasize that this statement should be considered as a necessary condition. An important property of the system (7.7), (7.8) is that, contrary to the equations of the ideal fluid approximation, it allows the possibility of imposing a *no-slip condition* at the body surface, i.e. the equality of not only the normal but also the tangential components of the fluid velocity and the body velocity at the body surface; a Newtonian viscous fluid admits a transverse transfer of momentum.

We will consider several instructive examples.

Example 1. The simplest example corresponds to Newton's original experiment (see Figure 7.1a). For the velocity distribution of the shear flow $u_1 = u(x_2)$, $u_2 = u_3 = 0$, the vector Navier–Stokes equation reduces to a single scalar equation,

$$\frac{d^2 u}{dx_2^2} = 0,$$

(7.19)

so that $u = Ax_2 + B$, where A and B are constants which we can obtain from the no-slip condition at the lower and upper plates,

$$u(0) = 0, \qquad u(h) = V.$$

(7.20)

Thus $B = 0$ and $A = V/h$, and we obtain (Figure 7.1b)

$$u = V \frac{x_2}{h}.$$

(7.21)

The viscous stress is constant across the whole layer:

$$\tau_{12} = \eta \frac{V}{h},$$

(7.22)

in full agreement with Newton's law (7.1).

Here there remains, however, one question (Barenblatt and Chernyi, 1963). We have automatically constructed the solution in a class of continuous functions with a continuous derivative. What is the physical nature of this restriction on the class of functions where the solution is sought? Why are solutions which have a discontinuity (Figure 7.1c) or a kink (Figure 7.1d) inappropriate?

Let us consider first a "candidate" solution having a kink (a discontinuity in the velocity derivative, Figure 7.1d). We can smooth, in an arbitrary monotonic way, the velocity distribution in the vicinity of a kink between $x_2 = x_{20} - \varepsilon$ and $x_2 = x_{20} + \varepsilon$, where $\varepsilon > 0$ is arbitrarily small (Figure 7.2a). The smoothed distribution does not satisfy equation (7.19); it satisfies a different equation,

$$\rho \frac{d^2 u}{dx_2^2} = \rho F(x_2),$$

(7.23)

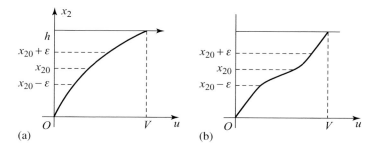

Figure 7.2 Smoothing of the velocity distributions shown in Figures 7.1:
(a) Figure 7.1d, (b) Figure 7.1a.

where the function $F(x_2)$ is a fictitious mass force obtained by differentiation of
the smoothed velocity distribution. The function $F(x_2)$ differs from zero only in
the interval $x_{20} - \varepsilon < x_2 < x_{20} + \varepsilon$. Integrating (7.23) from $x_{20} - \varepsilon$ to $x_{20} + \varepsilon$, we
obtain

$$v[(\partial_2 u)_{x_{20}+\varepsilon} - (\partial_2 u)_{x_{20}-\varepsilon}] = \int_{x_{20}-\varepsilon}^{x_{20}+\varepsilon} F(x_2)\, dx_2 \ . \tag{7.24}$$

Now let ε tend to zero. We see that, if the viscosity $v = \eta/\rho$ is different from zero,
the discontinuity of the velocity derivatives means the existence of a concentrated
force, the effect of which can be seen if we put a solid sheet at $x_2 = x_{20}$ and force
the sheet to move along the flow. In the case of an ideal fluid ($v = 0$) there is no
restriction on the discontinuity of the velocity derivative.

In the case where the "candidate" for the solution has a velocity discontinuity
at x_{20} (Figure 7.2b) we smooth the velocity distribution over the same interval,
$x_{20} - \varepsilon < x_2 < x_{20} + \varepsilon$, and come to the same equation, (7.23). Multiplying both
sides of (7.23) by $x_2 - x_{20}$ and integrating from $x_{20} - \varepsilon$ to $x_{20} + \varepsilon$, we obtain

$$v \int_{x_{20}-\varepsilon}^{x_{20}+\varepsilon} \left(\frac{\partial^2 u}{\partial x_2^2}\right)(x_2 - x_{20})\, dx_2 = v(x_2 - x_{20})(\partial_2 u)\Big|_{x_{20}-\varepsilon}^{x_{20}+\varepsilon} - v \int_{x_{20}-\varepsilon}^{x_{20}+\varepsilon} (\partial_2 u)\, dx_2$$

$$= O(\varepsilon) - v[u(x_{20} + \varepsilon) - u(x_{20} - \varepsilon)]$$

$$= \int_{x_{20}-\varepsilon}^{x_{20}+\varepsilon} (x_2 - x_{20})F(x_2)\, dx_2 \ . \tag{7.25}$$

Letting ε tend to zero, we obtain that if the function $u(x_2)$ has a discontinuity at
$x = x_{20}$ then this means that there is a concentrated couple (a vortex sheet), the
effects of which can be seen if at x_{20} we put two solid sheets, one over the other,
and move the sheets in opposite directions.

Requiring the solution to the Navier–Stokes equations to be smooth means physically that in the flow there must be no concentrated forces or couples.

Example 2. This example is concerned with the flow in a pipe of radius R where there is a pressure gradient $\partial_1 p$ along the pipe. For the shear flow established far from the entrance and exit to the pipe, where the only component of the velocity $u(r)$ is directed along the x_1 axis (here we use cylindrical coordinates r, θ, x_1), the Navier–Stokes equations reduce to

$$\frac{\eta}{r} \partial_r (r \partial_r u) = -\partial_1 p \,, \qquad \partial_r p = 0 \,, \qquad \partial_\theta p = 0 \,. \tag{7.26}$$

We note that, according to the second and third equations of (7.26), $\partial_1 p$ can depend only on x_1, whereas according to the first equation $\partial_1 p$ is equal to a function of the radius r. Therefore $\partial_1 p$ can only be a constant and is equal to $\Delta p / \ell$, where Δp is the pressure drop between the ends of the pipe and ℓ is its length, which is required to be much larger than the pipe radius R for the intermediate asymptotic region considered here to exist. Integrating the first equation in (7.26) and using the conditions $(\partial_r u)_{r=0} = 0$ (by symmetry) and $u = 0$ at $r = R$ (the no-slip condition at the pipe walls), we obtain

$$u = \frac{\Delta P}{\ell \eta} (R^2 - r^2) \,. \tag{7.27}$$

The total discharge, i.e. the fluid mass flowing through the pipe's cross-section in unit time, is equal to

$$Q = \int_0^R 2\pi r u \, dr = \frac{\pi \Delta P}{8 \ell \eta} R^4 \,. \tag{7.28}$$

Also, the mean velocity is of interest, i.e. the total discharge divided by the cross-section area:

$$\bar{u} = \frac{Q}{\pi R^2} = \frac{\Delta P}{8 \ell \eta} R^2 \,. \tag{7.29}$$

In engineering applications the dimensionless drag coefficient λ is often used:

$$\lambda = \frac{8 u_*^2}{\bar{u}^2} \,, \tag{7.30}$$

where $u_* = (\tau / \rho)^{1/2}$ and τ is the shear stress at the wall. As can easily be shown,

$$\tau = \frac{(\Delta P) \times \text{(cross-sectional area)}}{\text{(wetted area of the pipe surface)}} = \frac{\Delta P (\pi R^2)}{2 \pi R \ell} = \frac{\Delta P R}{2 \ell} \,,$$

so that $u_*^2 / \bar{u}^2 = 32 \nu^2 \ell / (\Delta P R^3)$. At the same time the Reynolds number Re is given by $\rho \bar{u} 2R / \eta = \Delta P R^3 / (4 \nu^2 \ell)$. Thus the expression for the drag coefficient in the pipe flow is

$$\lambda = \frac{64}{Re} \,. \tag{7.31}$$

Example 3. The half-space $x_2 > 0$, bounded by a flat plate at $x_2 = 0$, is filled with a viscous fluid at rest. At the initial moment $t = t_0$ the plate instantaneously starts to move with a constant speed V. A shear flow appears, and the velocity of this shear flow, $u_1 = u(x_2, t)$, depends only on the transverse coordinate x_2 and the time t; the velocity components u_2 and u_3 are equal to zero. As before, the continuity equation is satisfied automatically. The Navier–Stokes equation gives $\partial_2 p = \partial_3 p = 0$ and is reduced to the form

$$\partial_t u = -\frac{1}{\rho}\partial_1 p + v\partial_{22}^2 u . \tag{7.32}$$

It is clear that (cf. the previous example) $\partial_1 p$ can be a function only of time. At large distances from the plate the fluid remains at rest, therefore $\partial_1 p = 0$. We see that the Navier–Stokes equation is reduced in this case to the classical linear heat conduction equation,

$$\partial_t u = v\partial_{22}^2 u . \tag{7.33}$$

The initial and boundary conditions have the form

$$u(x_2, t_0) = 0 , \qquad u(0, t) = V , \qquad u(\infty, t) = 0 . \tag{7.34}$$

In this case we can consider the velocity dimension V as independent, so that

$$[u] = [V] , \qquad [x_2] = L , \qquad [t - t_0] = T . \tag{7.35}$$

Using the standard procedure of dimensional analysis presented in Chapter 2, we obtain

$$u = Vf\left(\frac{x_2}{\sqrt{v(t - t_0)}}\right) . \tag{7.36}$$

Substituting (7.36) into (7.33), we obtain an ordinary differential equation and boundary conditions:

$$\frac{d^2 f}{d\xi^2} + \tfrac{1}{2}\xi\frac{df}{d\xi} = 0 , \qquad f(0) = 1 , \qquad f(\infty) = 0 . \tag{7.37}$$

Here $\xi = x_2 / \sqrt{v(t - t_0)}$. Integration is simple, and we obtain eventually

$$u = \frac{2V}{\sqrt{\pi}} \int_{\frac{x_2}{2\sqrt{v(t-t_0)}}}^{\infty} e^{-z^2} dz . \tag{7.38}$$

This solution belongs to Lord Rayleigh. It looks simple, but it requires a discussion which promises to be instructive.

(1) The time t does not enter the basic equation explicitly. Therefore if we will start to reckon the time from a different moment, the equation does not change. This gives us the opportunity to consider as a governing parameter not the time

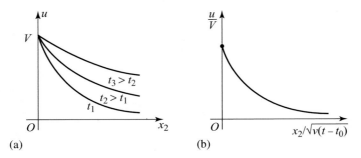

Figure 7.3 (a) The velocity distribution in the Rayleigh problem at different moments of time. (b) The universal velocity distribution in the same problem reduced coordinates.

t itself but $t - t_0$, which remains invariant with respect to a shift in the initial time, t_0.

(2) As stated above, the dimension of velocity in this problem could be considered as independent. Indeed, if we introduce an independent velocity unit (i.e. pass to the class $LMTV$) and vary only the velocity unit, the formulation of the problem (7.33), (7.34) will remain the same.

 In fact, this property and the property of invariance with respect to a shift in the initial time (item 1 above) are simple examples of the role of group invariance, which will be considered later in more detail.

(3) The solution (7.38) is self-similar. Introducing the velocity scale V and a time-dependent length scale $\sqrt{\nu(t - t_0)}$, we obtain a "frozen", i.e. time-independent, reduced velocity distribution: the velocity distributions at various moments can be obtained one from another by similarity transformations (Figure 7.3a). The frozen velocity distribution in reduced coordinates is presented in Figure 7.3b.

(4) The intermediate asymptotic character of the problem enables a significant simplification to be made, resulting in Rayleigh's solution (7.38).

 In fact, consider the motion of a plate of finite thickness h and length $\ell \gg h$ in a large tank whose size Λ is much larger than the length of the plate: $\Lambda \gg \ell$. The tank is filled with a fluid at rest. The motion of the plate begins at $t = t_0$; after a certain time δt the velocity of the plate reaches the value V and afterward remains constant until time t_1, when it stops. The distance of the plate from the tank boundary is of the order Λ and so is much larger than ℓ. We will consider the motion on one side of the plate only, invoking symmetry. The motion is considered at intermediate times $\delta t \ll t - t_0 \ll t_1 - t_0$. Also, it is considered at distances from the edges of the plate that are much larger than h. The transverse distances from the plate at which the motion will be considered are much less than ℓ, that is, $\ell \gg x_2$. Obviously, these intermediate asymptotic considerations, which led to the classic simple self-similar solution (7.38), were suggested to Lord Rayleigh by his intuition; however, it is instructive to trace the path of analysis directed by the intuition of great physicists.

7.4 The viscous dissipation of mechanical energy into heat

The total energy of a body occupying a region Ω at time t is equal to

$$\mathcal{E} = \int_{\Omega} \left(\rho c T + \rho \frac{u^2}{2} \right) d\omega , \tag{7.39}$$

where T is the temperature and c the heat capacity; thus

$$\partial_t \mathcal{E} = \int_{\Omega} (\rho c \partial_t T + \rho \mathbf{u} \cdot \partial_t \mathbf{u}) \, d\omega . \tag{7.40}$$

The time derivative of the temperature, $\partial_t T$, can be obtained from the heat conduction equation:

$$\rho c \partial_t T = -\nabla \cdot \mathbf{q} + Q , \tag{7.41}$$

where $\mathbf{q}$ is the heat flux and Q is the rate of heat generation in the volume. The velocity time derivative $\partial_t \mathbf{u}$ can be obtained from the momentum balance equation:

$$\partial_t \mathbf{u} = -(\mathbf{u} \cdot \nabla)\mathbf{u} - \frac{1}{\rho} \nabla p + \frac{1}{\rho} \nabla \boldsymbol{\tau} . \tag{7.42}$$

Substituting (7.41) and (7.42) into (7.40), we arrive at the equation

$$\partial_t \mathcal{E} = \int_{\Omega} \nabla \cdot (-\mathbf{q}) \, d\omega - \int_{\Omega} \rho \mathbf{u} \cdot (\mathbf{u} \cdot \nabla)\mathbf{u} \, d\omega - \int_{\Omega} \mathbf{u} \cdot \nabla p \, d\omega + \int_{\Omega} (\mathbf{u} \cdot \nabla \boldsymbol{\tau}) \, d\omega . \tag{7.43}$$

We now transform the last three terms on the right-hand side of (7.43) respectively, bearing in mind that $\nabla \cdot \mathbf{u} = 0$ (the continuity equation):

$$-\int_{\Omega} \rho \mathbf{u} \cdot (\mathbf{u} \cdot \nabla)\mathbf{u} \, d\omega = -\int_{\Omega} \rho \nabla \cdot \left(\mathbf{u} \frac{u^2}{2} \right) d\omega ,$$

$$-\int_{\Omega} \mathbf{u} \cdot \nabla p \, d\omega = -\int_{\Omega} \nabla \cdot (p\mathbf{u}) \, d\omega ,$$

$$\int_{\Omega} (\mathbf{u} \cdot \nabla \boldsymbol{\tau}) \, d\omega = \int_{\Omega} u_\gamma \partial_\beta \tau_{\gamma\beta} \, d\omega$$

$$= \int_{\Omega} \partial_\beta u_\gamma \, \tau_{\gamma\beta} \, d\omega - \int_{\Omega} \tau_{\gamma\beta} \partial_\beta u_\gamma \, d\omega$$

$$= \int_{\Omega} \nabla \cdot (\mathbf{u} \cdot \boldsymbol{\tau}) \, d\omega - \int_{\Omega} \boldsymbol{\tau} \cdot \nabla \cdot \mathbf{u} \, d\omega .$$

Finally we obtain

$$\partial_t \mathcal{E} = \int_\Omega \nabla \cdot \mathbf{G} \, d\omega + \int_\Omega (Q - \boldsymbol{\tau} \cdot \nabla \cdot \mathbf{u}) \, d\omega \qquad (7.44)$$

where

$$\mathbf{G} = -\mathbf{q} - \mathbf{u}\frac{\mathbf{u}^2}{2} - p\mathbf{u} + \mathbf{u} \cdot \boldsymbol{\tau} . \qquad (7.45)$$

The first volume integral on the right-hand side of (7.44) can be transformed into the surface integral

$$\int_\Omega \mathbf{G} \cdot \mathbf{n} \, d\Sigma = \int_{\partial\Omega} \left(-\mathbf{q} - \mathbf{u}\frac{\mathbf{u}^2}{2} - p\mathbf{u} + \mathbf{u} \cdot \boldsymbol{\tau} \right) \cdot \mathbf{n} \, d\Sigma, \qquad (7.46)$$

which represents the flux of the total energy via the surface $\partial\Omega$ of the region Ω. The second volume integral represents the generation rate of the total energy in the region Ω. In absence of endo- or exothermic reactions the total energy is conserved so, owing to the arbitrariness of Ω, the integrand must be equal to zero, and the rate of heat generation in the volume Ω is given by

$$Q = \boldsymbol{\tau} \cdot \mathbf{D} \qquad (7.47)$$

because $\boldsymbol{\tau} \cdot \nabla \cdot \mathbf{u} = \boldsymbol{\tau} \cdot (\nabla \cdot \mathbf{u} + (\nabla \cdot \mathbf{u})^{\mathrm{T}})/2 = \boldsymbol{\tau} \cdot \mathbf{D}$. For a Newtonian viscous fluid of constant density, $\boldsymbol{\tau} = 2\eta\mathbf{D}$ (see equation (7.5)); therefore

$$Q = 2\eta\mathbf{D} \cdot \mathbf{D} = 2\eta\mathbf{D}^2 = \frac{\eta}{2} (\partial_\alpha u_\beta + \partial_\beta u_\alpha)^2 . \qquad (7.48)$$

8

The Newtonian viscous fluid approximation.
Applications: the boundary layer

8.1 The drag on a moving wing. Friedrichs' example

We return now to the problem of the steady flow around a thin weakly inclined wing. Using the ideal incompressible fluid approximation we were able to obtain the lift force acting on the wing. As far as the drag is concerned we found that the ideal incompressible fluid approximation is insufficient: the drag obtained in this approximation appeared to be equal to zero (the so-called D'Alembert paradox). So, we will use the Newtonian viscous fluid model to calculate the drag; for a thin wing under a small angle of attack the basic contribution to the drag is given by the viscous stresses on the wing. We remind the reader that motion with not too large a velocity is considered, up to 300–400 km/hour, which was the speed of many military aircraft before and during World War II.

We introduce the dimensionless variables

$$\boldsymbol{\xi} = \frac{\mathbf{x}}{L}, \qquad \mathbf{v} = \frac{\mathbf{u}}{U}, \qquad P = \frac{p}{\rho U^2}, \qquad (8.1)$$

where, we remind the reader, L is the "chord" of the wing and U is its velocity (see Figure 4.2b). In the dimensionless variables (8.1), the Navier–Stokes equation for the steady flow under consideration takes the form

$$(\mathbf{v} \cdot \nabla)\mathbf{v} = -\nabla P + \varepsilon \Delta \mathbf{v} . \qquad (8.2)$$

Here the operators ∇ and Δ are taken in the dimensionless variables $\xi_i = x_i/L$, and the parameter ε is given by

$$\varepsilon = \frac{1}{Re}, \qquad Re = \frac{UL}{\nu} . \qquad (8.3)$$

The parameter Re is a basic similarity parameter, the Reynolds number.

Let us evaluate the parameter ε. For the period of World War II, at the altitude fighters could then reach, the values of the parameters in (8.3) can be taken as $\nu = 0.15\,\mathrm{cm}^2/\mathrm{s}$, $U \sim 100\,\mathrm{m/s} = 10^4\,\mathrm{cm/s}$ and $L \sim 1\,\mathrm{m} = 10^2\,\mathrm{cm}$, so that the parameter ε is of the order of 10^{-7}, one ten-millionth (in fact even less). However, the term with this small parameter cannot be omitted otherwise we would merely obtain the Euler equation and with it the D'Alembert paradox: no drag.

Two points should be mentioned. First, no analytic solutions of the Navier–Stokes equations for flows around bodies have been obtained. Second, as we demonstrated in Chapter 4, the effects of viscosity are concentrated in a thin layer at the wing's boundary, the boundary layer.

To gain an understanding of the steps that follow, we consider an illuminating mathematical example for which everything can be calculated explicitly. This example was proposed by the American mathematician K. O. Friedrichs (1966), and it has played a substantial role in clarifying the mathematical side of the problem.

Consider the second-order differential equation with a small parameter ε at the highest derivative,

$$\varepsilon \frac{d^2 f}{dx^2} + \frac{df}{dx} = a , \qquad (8.4)$$

where $0 < a < 1$ is also a parameter, under the boundary conditions

$$f(0) = 0 , \qquad f(1) = 1 . \qquad (8.5)$$

For this case a simple solution of the problem is available:

$$f(x) = ax + (1 - a)(1 - e^{-x/\varepsilon})(1 - e^{-1/\varepsilon}) . \qquad (8.6)$$

A simple analysis shows that if the value $\varepsilon = 10^{-7}$ is taken, then, already for x equal to or more than one-millionth, the solution (8.6) is practically indistinguishable from the function

$$f_0 = ax + (1 - a) . \qquad (8.7)$$

However, this is the very solution obtained if we drop the term with the higher-order derivative in (8.4) and satisfy the second boundary condition (8.5). Although the exact solution (8.6) is very close to the function $f_0(x)$ if one starts from values of x very close to zero, the approximate solution (8.7) cannot be used if we are interested in the value of $\varepsilon(df/dx)$ at $x = 0$, which is an analogue of the shear stress at the boundary. For this quantity the exact solution (8.6) gives

$$\varepsilon \left(\frac{df}{dx} \right)_{x=0} = \frac{1 - a}{1 - e^{-1/\varepsilon}} . \qquad (8.8)$$

To any reasonable degree of accuracy this value is equal to $1 - a$. Thus the approximate solution f_0 is useless for calculating the value $\varepsilon(df/dx)_{x=0}$ in which we are interested.

Therefore, to calculate the quantity $\varepsilon(df/dx)_{x=0}$ we will use an approximate approach which demonstrates the basic idea of the theory of the boundary layer; this was the intention of Friedrichs.

To investigate the behaviour of the solution in a small vicinity of $x = 0$, we introduce the "stretched" coordinate

$$X = x/\varepsilon . \tag{8.9}$$

This stretched coordinate is introduced in such a way that the two first terms in equation (8.4) are of the same order of magnitude. Equation (8.4) in these new variables $X = x/\varepsilon$, $\mathcal{F}(X) = f(x)$ takes the form

$$\frac{d^2\mathcal{F}}{dX^2} + \frac{d\mathcal{F}}{dX} = a\varepsilon . \tag{8.10}$$

As $\varepsilon \to 0$ the term on the right-hand side becomes negligibly small in comparison with the terms on the left-hand side, and we can neglect it. Integrating the equation $d^2\mathcal{F}/dX^2 + d\mathcal{F}/dX = 0$ thus obtained and imposing the condition $\mathcal{F}(0) = 0$, we obtain

$$\mathcal{F}(X) = C(1 - e^{-X}) . \tag{8.11}$$

The constant C is determined in the following way. We require that the approximate solution $f_0(x)$, which is valid outside a small neighborhood of $x = 0$, matches continuously the approximate solution (8.11), which is valid in a small neighborhood of $x = 0$, at a point $x = \delta$, say. In this case we have

$$C(1 - e^{-\delta/\varepsilon}) = a\delta + (1 - a) . \tag{8.12}$$

Now we let δ tend to zero together with the small parameter ε, but more slowly, so that $\delta/\varepsilon \to \infty$. In the limit, the relation (8.12) gives $C = 1-a$, so that the approximate solution in the neighborhood of $x = 0$ ultimately takes the form

$$\mathcal{F}(X) = f(x) = (1 - a)(1 - e^{-x/\varepsilon}) . \tag{8.13}$$

Differentiating (8.12), we find

$$\varepsilon\left(\frac{df}{dx}\right)_{x=0} = 1 - a . \tag{8.14}$$

This approximate value practically coincides with the exact value (8.8). (The difference between $1 - e^{-1/\varepsilon}$ and 1 cannot play a role.)

In summary, to construct an adequate approximation to the solution to equation (8.4), which contains a small parameter at the highest derivative under the conditions (8.5) prescribed on both sides of the interval, we used the following special procedure.

Step A. We ignored the term with the small parameter at the highest derivative. We obtained a first-order equation, a simplified *outer-approximation equation*, which was integrated with a given condition at the end $x = 1$ of the interval. The solution obtained did not satisfy the condition at the origin $x = 0$ of the interval, although it practically coincided with the exact solution outside a small vicinity of $x = 0$, the boundary layer, where the variation of the solution is fast.

Step B. Approximating the solution in the vicinity of $x = 0$, we stretched the coordinate x, so that in the boundary layer the term with the small parameter at the second derivative had the same order of magnitude as the other term on the left-hand side of (8.4).

The equation obtained, the *inner-approximation equation*, was integrated for the given initial value at the left-hand end $x = 0$ of the interval. It is very important that *as the stretched variable tends to infinity the inner-approximation solution becomes equal to the value of the outer approximation at zero.*

8.2 Model of the boundary layer at a thin weakly inclined wing of infinite span

The model of an aircraft wing with infinite span was discussed in Chapter 4 and will be of basic importance here too. The Navier–Stokes equation and the equation of continuity for steady motion around the wing take the forms

$$
\begin{aligned}
u_1 \partial_1 u_1 + u_2 \partial_2 u_2 &= -\frac{1}{\rho}\, \partial_1 p + \nu \left(\frac{\partial^2 u_1}{\partial x_1^2} + \frac{\partial^2 u_1}{\partial x_2^2} \right), \\
u_1 \partial_1 u_2 + u_2 \partial_2 u_2 &= -\frac{1}{\rho}\, \partial_2 p + \nu \left(\frac{\partial^2 u_2}{\partial x_1^2} + \frac{\partial^2 u_2}{\partial x_2^2} \right), \\
\partial_1 u_1 + \partial_2 u_2 &= 0\,,
\end{aligned}
\tag{8.15}
$$

where ν is the kinematic viscosity. In contrast with Friedrichs' example (see equation (8.4)), the exact solution of the Navier–Stokes equation for this problem is unknown; there exists not a single solution to the problem of the flow around bodies, even the simplest. Therefore in our model we will use the the *Prandtl hypothesis*, based on experimental fact: the effects of viscosity (i.e. the terms containing the viscosity ν in the Navier–Stokes equations) govern the flow in only a thin boundary layer adjacent to the wing surface. On the basis of this hypothesis we will obtain approximate equations for motion in the boundary layer (the *boundary layer equations*). The solution to these equations makes it possible to calculate the shear stress at the wing surface and the drag force.

It should be emphasized that up to the present time a rigorous derivation of the boundary layer equations does not exist. As before, obtaining these equations remains a memorial to the intuition of a genius, the great German physical scientist L. Prandtl (Prandtl, 1905), who presented them in 1904.[1] We described earlier Friedrichs' example, for which the Prandtl procedure is transparent and could be made rigorous. This example was presented above to give the reader the chance to gain the necessary intuition to follow a *non-rigorous* derivation of the boundary layer equations.

So, we reduce equations (8.15) to dimensionless form using the variables (8.1):

$$v_1\partial_1 v_1 + v_2\partial_2 v_1 = -\partial_1 P + \varepsilon(\partial_{11}^2 v_1 + \partial_{22}^2 v_1),$$
$$v_1\partial_1 v_2 + v_2\partial_2 v_2 = -\partial_2 P + \varepsilon(\partial_{11}^2 v_2 + \partial_{22}^2 v_2),$$
$$\partial_1 v_1 + \partial_2 v_2 = 0.$$

(8.16)

Here, the symbol ∂_i denotes the derivative of the dimensionless variable ξ_i; the v_i are the components of the dimensionless vector $\mathbf{v}$ (see (8.1));[2] as before, $\varepsilon = 1/Re$.

Dropping, outside the boundary layer, the terms of the equations containing the viscosity, i.e. those involving the small parameter ε, we obtain an analogue of the outer approximation in Friedrichs' example $df_0/dx = a$: the Euler equations for ideal incompressible fluid flow, considered before in Chapter 4 where we presented their solution for the flow around a wing of infinite span. According to that solution the tangential velocity component v_0 at the wing surface is different from zero. The function $v_0(s)$, where s is the curvilinear coordinate reckoned from the critical point where the velocity is equal to zero, can be considered as known. The analogue of $v_0(s)$ in Friedrichs' example is the value f_0 at $x = 0$, which is different from zero and equal to $1 - a$. Consider now the equations of motion in the thin boundary layer. In line with Prandtl's hypothesis we introduce in the boundary layer a local system of Cartesian coordinates such that x_1 is directed along the wing surface and x_2 is directed along the normal to the wing surface. The radius of curvature of the wing surface is much larger than the thickness of the boundary layer everywhere, except in the vicinity of the leading edge. Therefore we can neglect the curvature of the coordinate s.

Now we stretch the normal coordinate ξ_2, leaving the longitudinal coordinate invariant, and introduce new variables:

$$X_1 = \xi_1, \qquad X_2 = \frac{\xi_2}{\sqrt{\varepsilon}}, \qquad U_1 = v_1, \qquad U_2 = \frac{v_2}{\sqrt{\varepsilon}}.$$

(8.17)

---

[1] The first presentation of these equations was delivered by the 29-year-old Prandtl, by invitation of F. Klein, at the Third International Congress of Mathematicians in Heidelberg in 1904. Barely ten people were present at Prandtl's lecture; apparently no one was able to appreciate the importance of the event. Prandtl was so exhausted describing his ideas that after the lecture he fainted.

[2] The reader will note the similarity between the conventional symbols v for velocity component and ν for viscosity.

The factors $\sqrt{\varepsilon}$ in the denominators of X_2 and U_2 are introduced so that the viscous term $\varepsilon\partial^2_{22}v_1$ in the first equation (8.16) remains of the same order as the basic term in the first equation of (8.16), $v_1\partial_1 v_1 = U_1\partial_1 U_1$ and both terms in the continuity equation have the same order of magnitude. After the transition to the variables (8.17), the equations (8.16) assume the form

$$U_1\partial_1 U_1 + U_2\partial_2 U_1 = -\partial_1 P + \varepsilon\partial^2_{11}U_1 + \partial^2_{22}U_1 \, ,$$
$$\partial_2 P = O(\sqrt{\varepsilon}) \, , \tag{8.18}$$
$$\partial_1 U_1 + \partial_2 U_2 = 0 \, .$$

Note that in equations (8.18) the symbols ∂_1 and ∂_2 correspond to differentiation by X_1 and X_2. We shall neglect in equations (8.18) terms of the order of $\sqrt{\varepsilon}$ and ε, in analogy to neglecting the term $a\varepsilon$ in Friedrichs' example. The equations for the flow in the boundary layer now take the form

$$U_1\partial_1 U_1 + U_2\partial_2 U_1 = -\partial_1 P + \partial^2_{22}U_1 \, ,$$
$$\partial_1 U_1 + \partial_2 U_2 = 0 \, , \tag{8.19}$$
$$\partial_2 P = 0 \, .$$

The next step is matching the solution to these equations at $X_2 = \infty$ with the solution of the outer approximation (ideal fluid flow) at the wing surface; this is the analogue of the condition $F(\infty) = f_0(1) = 1 - a$ in Friedrichs' example. Indeed, in the boundary layer we have $\partial_2 P = 0$, so that the function P does not depend on X_2 inside the boundary layer. Therefore (since the boundary layer is thin) we can use the expression for P from the solution to the problem of ideal fluid flow at the wing surface. For such an ideal fluid flow the Bernoulli integral is valid and can be written in the form

$$P + \tfrac{1}{2} U_0^2(X_1) = \mathrm{const} \, , \tag{8.20}$$

where $U_0(X_1)$ is a function that can be assumed to be known from the solution to the ideal fluid flow problem. Therefore $\partial_1 P$, which enters the first equation in (8.19), can be replaced by a known function, $-U_0(X_1)\partial_1 U_0(X_1)$. We finally arrive at the boundary layer equations in their ultimate form:

$$U_1\partial_1 U_1 + U_2\partial_2 U_1 = U_0(X_1)\partial_1 U_0(X_1) + \partial^2_{11}U_1 \, ,$$
$$\partial_1 U_1 + \partial_2 U_2 = 0 \, . \tag{8.21}$$

Equations (8.21) are complemented by the following boundary conditions. The first is the natural no-slip condition at the bottom of the boundary layer (it is the analogue of the condition $\mathcal{F}(0) = 0$ in Friedrichs' example); this condition implies that

$$U_1(X_1, 0) = 0 \, , \qquad U_2(X_1, 0) = 0 \, . \tag{8.22}$$

The next condition prescribes the longitudinal velocity at a certain cross-section of the boundary layer $X_1 = X_{10}$:

$$U_1(X_{10}, X_2) = U_1^0(X_2) \,. \tag{8.23}$$

In fact, in the boundary layer equations (8.21) the first derivative $\partial_1 U_1$ enters but there is no second derivative $\partial_{11}^2 U_1$. This means that the transfer of the perturbation back along the flow in the boundary layer approximation, present in the full Navier–Stokes equation, is missing, so the longitudinal component of the velocity at the entrance to the boundary layer should be prescribed.

The most significant condition is that of matching the longitudinal velocity component at $X_2 = \infty$, at the top of the boundary layer, with the longitudinal velocity component obtained from the solution for the flow around the wing in the ideal incompressible fluid approximation at the wing surface:

$$U_1(X_1, \infty) = U_0(X_1) \,. \tag{8.24}$$

This is the analogue of the condition $\mathcal{F}(\infty) = f_0(0) = 1 - a$ in Friedrichs' example. The analogous condition for the transverse velocity component U_2 is not needed: in the boundary layer equations the term $\partial_{22}^2 U_2$ is missing, so that for U_2 only one of the conditions (8.22) is necessary.

Now, *after all these arguments, nothing prevents us from returning, in equations* (8.21) *and the boundary conditions* (8.22)–(8.24), *to the original dimensional quantities* x_1, x_2, u_1, u_2, v:

$$u_1 \partial_1 u_1 + u_2 \partial_2 u_1 = u_0 \partial_1 u_0 + v \partial_{22}^2 u_1 \,,$$
$$\partial_1 u_1 + \partial_2 u_2 = 0 \,, \tag{8.25}$$

$$u_1(x_2, 0) = 0 \,, \qquad u_2(x_1, 0) = 0$$
$$u_1(x_{10}, x_2) = u_0(x_2) \,, \qquad u_1(x_1, \infty) = u_0(x_1) \,. \tag{8.26}$$

We note, that in traditional derivations of the boundary layer equations, as found in major textbooks, the last condition would appear to be controversial. The crucial point in the derivation of the boundary layer equations is the thinness of the layer, nevertheless, the boundary condition at the outer boundary of the layer is taken to be at $x_2 = \infty$. Here, we have seen how this condition appears.

8.3 The boundary layer on a flat plate

An instructive example of a solution of the boundary layer equations is the intermediate asymptotic solution of the flow around a flat plate, obtained by Prandtl's student H. Blasius (Blasius, 1908) (see Figure 8.1). We remind the reader that there

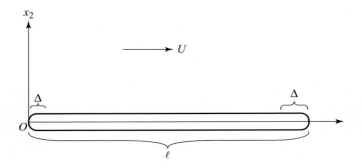

Figure 8.1 A flat plate in a uniform flow.

exists not a single analytic solution to the viscous flow around bodies, even for a flat plate; therefore this intermediate-asymptotic analytic solution is of especial importance.

So, consider a flat plate of thickness h and length ℓ in a uniform flow with velocity U. We will consider, in fact, the flow at distances from the leading and trailing edges that are much larger than h but less than ℓ.

In the intermediate region under consideration the uniform flow with velocity U remains undisturbed in the ideal incompressible fluid approximation. Therefore $u_0 = u^0 = U$, $\partial_1 P = 0$, so that the basic system of equations and boundary conditions assumes the form

$$u_1 \partial_1 u_1 + u_2 \partial_2 u_1 = \nu \partial^2_{22} u \,,$$
$$\partial_1 u_1 + \partial_2 u_2 = 0 \,,$$
$$u_1(x_1, 0) = u_2(x_1, 0) = 0 \,,$$
$$u_1(0, x_2) = U, \qquad u_1(x_1, \infty) = U \,. \tag{8.27}$$

In obtaining a solution to this problem we will use dimensional analysis, which in this case is non-trivial in many respects and therefore particularly instructive. According to the traditional approach the unknown quantities u_1 and u_2 depend on the governing parameters x_1, x_2, U and ν. The dimensions of u_1, u_2 and the governing parameters are as follows:

$$[u_1] = [u_2] = LT^{-1} \,, \qquad [U] = LT^{-1} \,, \qquad [\nu] = L^2 T^{-1} \,, \qquad [x_1] = [x_2] = L \,.$$

Dimensional analysis gives

$$u_1 = U\Phi_1(\xi_1, \xi_2) \,, \qquad u_2 = U\Phi_2(\xi_1, \xi_2) \,,$$
$$\xi_1 = \frac{U x_1}{\nu} \,, \qquad \xi_2 = \frac{U x_2}{\nu} \,. \tag{8.28}$$

Substituting (8.28) into (8.27) we obtain (the derivatives ∂_1 and ∂_2 are assumed to be performed with respect to ξ_1 and ξ_2)

$$\Phi_1\partial_1\Phi_1 + \Phi_2\partial_2\Phi_1 = \partial^2_{22}\Phi_1\,,$$
$$\partial_1\Phi_1 + \partial_2\Phi_2 = 0\,, \tag{8.29}$$
$$\Phi_1(\xi_1,0) = \Phi_2(\xi_1,0) = 0\,,$$
$$\Phi_1(\xi_1,\infty) = 1\,, \qquad \Phi_1(0,\xi_2) = 1\,.$$

This result appears disappointing: the resulting system repeats the original system (8.27) if the coefficients ν and U are put equal to 1.

We note now an essential point: if $\Phi_1(\xi_1,\xi_2)$ and $\Phi_2(\xi_1,\xi_2)$ constitute the solution to the problem (8.29) then the functions

$$\psi_1(\xi_1,\xi_2,\alpha) = \Phi_1(\alpha^2\xi_1,\alpha\xi_2)\,,$$
$$\psi_2(\xi_1,\xi_2,\alpha) = \alpha\Phi_2(\alpha^2\xi_1,\alpha\xi_2) \tag{8.30}$$

also satisfy the equations and the boundary conditions (8.29) for any positive values of α. This property is a special example of invariance with respect to a transformation group. Indeed, the transformation (8.30) is a one-parameter group of transformations, since:

(1) among the transformations (8.30) there is an identity, corresponding to $\alpha = 1$;
(2) to each transformation (8.30) there corresponds an inverse having the parameter $\beta = 1/\alpha$, so that the transformed quantities return to the originals;
(3) sequentially applying the transformations corresponding to the parameters α and β, we obtain the transformed quantities corresponding to the parameter $\alpha\beta$ contained in the same set.

Now, it can be proved that the solution to the problems (8.27) and, consequently, (8.29) exists and is unique. Therefore $\Phi_1 = \Psi_1$ and $\Phi_2 = \Psi_2$ at any $\alpha > 0$. That means that the functions $\Phi_1(\xi_1,\xi_2)$ and $\Phi_2(\xi_1,\xi_2)$ satisfy the following relations:

$$\Phi_1(\xi_1,\xi_2) = \Phi_1(\alpha^2\xi_1,\alpha\xi_2)\,, \qquad \Phi_2(\xi_1,\xi_2) = \alpha\Phi_2(\alpha^2\xi_1,\alpha\xi_2)\,. \tag{8.31}$$

It is important that, having derived relations (8.31), we may set α in these relations equal to any positive quantity, in particular $\alpha = 1/\sqrt{\xi_1}$. We then obtain

$$\Phi_1(\xi_1,\xi_2) = \Phi_1\left(1,\frac{\xi_2}{\sqrt{\xi_1}}\right) = f_1\left(\frac{\xi_2}{\sqrt{\xi_1}}\right) = f_1\left(\frac{x_2}{\sqrt{\nu x_1/U}}\right),$$

$$\Phi_2(\xi_1,\xi_2) = \left(\frac{1}{\sqrt{\xi_1}}\right)\Phi_2\left(1,\frac{\xi_2}{\sqrt{\xi_1}}\right) = \frac{1}{\sqrt{\xi_1}}\,f_2\left(\frac{\xi_2}{\sqrt{\xi_1}}\right) = \sqrt{\frac{\nu}{Ux_1}}\,f_2\left(\frac{x_2}{\sqrt{\nu x_1/U}}\right).$$
$$\tag{8.32}$$

We have established finally the *self-similarity* of the solution. It is instructive, however, that in this case the self-similarity was obtained using not only classical dimensional analysis but also the invariance of the solution under the operations of an additional transformation group.

This invariance allows a certain formal procedure of dimensional analysis, and this device can sometimes be helpful. Let us measure the lengths in the longitudinal x_1 and transverse x_2 directions using different units, i.e. we introduce *two length dimensions*, L_1 and L_2. It is possible to do this for the boundary layer equations, but not for the full Navier–Stokes equations because the term $\partial_{22}^2 u_1$ enters these equations together with the term $\partial_{11}^2 u_1$. So, if we measured x_1 and x_2 in different units then these terms would have different dimensions, and that is impossible.

So, in the case of the boundary layer equations it is possible to achieve dimensional homogeneity by taking the dimensions of the governing and governed parameters as follows:

$$[u_1] = [U] = L_1 T^{-1} , \quad [\nu] = L_2^2 T^{-1} , \quad [u_2] = L_2 T^{-1} , \quad [x_1] = L_1 , \quad [x_2] = L_2 .$$
(8.33)

In this case not two but *three* governing parameters will have independent dimensions. The only dimensionless parameter will be

$$\zeta = \frac{x_2}{\sqrt{\nu x_1 / U}} .$$

From this follow the self-similarity relations (8.32).

Now we continue the solution of equations (8.23). Introduce a new function

$$\phi(\zeta) = \int_0^\zeta f_1(\zeta)\, d\zeta ,$$
(8.34)

and substitute (8.34) into the continuity equation, the second equation in (8.27). By integration of the resulting ordinary differential equation and using the relation (8.34), we obtain a relation between f_2 and f_1:

$$f_2 = \tfrac{1}{2}(\zeta \phi' - \phi) .$$
(8.35)

Furthermore, the momentum balance equation for the boundary layer (the first equation in (8.27)) leads to an ordinary differential equation of the third order,

$$\phi \phi'' + 2\phi''' = 0 .$$
(8.36)

The boundary conditions we require follow directly from the boundary conditions presented in (8.27):

$$\phi(0) = \phi'(0) = 0 , \quad \phi'(\infty) = 1 .$$
(8.37)

Thus the boundary layer problem under consideration is reduced to a boundary value problem for an ordinary differential equation of the third order with boundary conditions at each end of the infinite interval $(0, \infty)$.

It is remarkable (Töpfer, 1912) that equation (8.36) and the first two boundary conditions at $\zeta = 0$ are also invariant with respect to a transformation group (see above). Indeed, it is easy to check that if $\phi(\zeta)$ is a solution to equation (8.36) satisfying the conditions $\phi(0) = \phi'(0) = 0$ then, for any $\alpha > 0$, the function $\alpha\phi(\alpha\zeta)$ satisfies the same equation for the same boundary conditions. This property allows the replacement of the more difficult solution of the boundary value problem with the solution to the Cauchy problem for equation (8.36) satisfying the initial data

$$\phi_0(0) = \phi_0'(0) = 0 , \qquad \phi_0''(0) = 1 . \tag{8.38}$$

This last solution is easy to compute. We take the solution in the form $\phi(\zeta) = \alpha\phi_0(\alpha\zeta)$. For the solution $\phi_0(\zeta)$, the derivative at infinity is, as an easy calculation shows, $\phi_0'(\infty) = 2.086$. Therefore taking $\alpha = 1/\sqrt{2.086} = 0.6925$ satisfies all the conditions of (8.37).

The solution obtained just now allows us to determine the drag on the plate. Indeed, in the intermediate interval $\Delta < x_1 < \ell - \Delta$ the tangential stress at the surface of the plate is

$$(\tau_{12})_{x_2=0} = \eta(\partial_2 u)_{x_2=0} = \eta U \sqrt{\frac{U}{\nu x_1}} \, f_1'(0)$$

$$= \frac{\eta U^{3/2}}{\sqrt{\nu x_1}} \, \alpha^2 \phi''(0) = 0.332 \, \eta U^{3/2} (\nu x_1)^{-1/2} . \tag{8.39}$$

Neglecting the contribution of the drag at the plate edges $0 < x_1 < \Delta$, $\ell - \Delta < x_1 < \ell$, we obtain, for the drag force $\mathcal{F}$,

$$\mathcal{F} = 2 \int_0^\ell (\tau_{12})_{x_2=0} \, dx_1 = 1.328 \, \rho (\nu \ell U^3)^{1/2} . \tag{8.40}$$

Introducing the dimensionless parameter $\Pi = \mathcal{F}/(\rho U^2 \ell)$ corresponding to the drag $\mathcal{F}$, we get

$$\Pi = \Phi(Re) = \frac{1.328}{\sqrt{Re}} , \qquad Re = \frac{U\ell}{\nu} . \tag{8.41}$$

Several comments are in order. First, the situations at the leading edge, $0 < x_1 < \Delta$, and the trailing edge, $\ell - \Delta < x_1 < \ell$, of the plate should be considered separately. At the leading edge there is a thin layer at the boundary of the plate where the viscous effects are concentrated. However, the boundary layer equations (8.27) cannot be used there because the radius of the curvature of the plate surface is small and so the derivative $\partial_{11}^2 u_1$ cannot be neglected. As experiments show, at

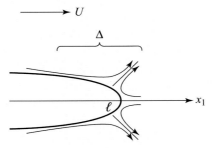

Figure 8.2 Flow separation at the trailing edge of a plate.

the end of this small region the distortion of the uniform flow becomes negligible: the flow at the entrance to the intermediate region under consideration can be taken as uniform and the condition $u_1(0, x_2) = U$ can be used.

The situation at the trailing edge is different. In this region, $\ell - \Delta < x_1 < \ell$, we have flow separation (see Figure 8.2). Owing to viscosity effects the energy of the fluid particles is insufficient to carry them against the gradient of pressure, which increases in the trailing part of the plate, so the fluid particles turn away from the surface layer and a flow appears in the opposite direction. Mathematically this means that the solution to the boundary layer equations ceases to exist. Nothing bad happens to the solution to the Navier–Stokes equations; however, the Prandtl hypothesis, which was the basis of the boundary layer approximation, is no longer valid.

Our assumption is in fact that, owing to the thinness of the plate, the size Δ of the trailing edge region is small and so it does not contribute to the drag $\mathcal{F}$ substantially.

The second comment is as follows. Naive application of dimensional analysis to the problem of the flow around the plate suggests the following steps. We accept as the governing parameters the flow velocity U, the length of the plate ℓ, the density ρ and the viscosity ν of the fluid. Then the dimensionless parameter $\Pi = \mathcal{F}/(\rho U^2 \ell)$ appears to be a function of Re only: $\Pi = \Phi(Re)$. However, Re is very large, of the order of 10^7 even for old aircraft. It seems natural therefore to replace $\Phi(Re)$ by a constant equal to $\Phi(\infty)$ and to obtain thereby the following expression for the drag:

$$\mathcal{F} = \text{const} \times \rho U^2 \ell \,, \tag{8.42}$$

which contradicts (8.41). The reason for this contradiction is that we have tacitly assumed that as $Re \to \infty$ the function $\Phi(Re)$ tends to a finite non-zero limit, which is not in fact correct: at large Re the function $\Phi(Re)$ has a power-type asymptotic $\Phi(Re) = 1.328 \, Re^{-1/2}$, so the limiting value of $\Phi(Re)$ as $Re \to \infty$ is zero. However, we are in fact not interested in the limiting value but rather in the asymptotics for large, though finite, values of *Re*. So, using the asymptotic result $\Phi(Re) = 1.328/\sqrt{Re}$, we obtain the correct expression (8.40) for the drag. This is in fact a simple example of "incomplete similarity", which will be considered in detail in the next chapter.

In conclusion, we reiterate that the solution for viscous flow around a flat plate was obtained above in the boundary layer approximation. Note that the corresponding solution for the Navier–Stokes equations has not yet been obtained. In Volume II of the posthumously published collected works of N. E. Kochin there is a paper which contains a claim for such a solution. This solution is incorrect but Kochin, who was never mistaken in his published works, was not responsible for this publication. As was mentioned in the footnote, the article was written by A. A. Dorodnitsyn, who used unfinished handwritten notes found in Kochin's papers after his death.

9

Advanced similarity methods: complete and incomplete similarity

9.1 Examples

We will start with two fundamental examples which illustrate the possibility of a useful modification of dimensional analysis in obtaining scaling laws. A more detailed presentation of the material in this chapter can be found in the author's books (Barenblatt, 1996, 2003). The remarkable book by N. D. Goldenfeld (Goldenfeld, 1992) is recommended for learning the connection between complete, and incomplete, similarity and the renormalization group.

The first example is a geometric one. It concerns *fractal* geometric objects. The general concept of a fractal object was introduced by the late French–American mathematician Benoit Mandelbrot (1975, 1977). Mandelbrot demonstrated with many examples that such objects can be widely used in the mathematical modeling of natural phenomena and in engineering.

First let us take a circle, and inscribe in it regular polygons with a growing number of sides n (Figure 9.1a). The perimeter L_n of the inscribed n-gon with side length η obviously depends only on the diameter of the circle d and the side length η; the number of sides of the n-gon is determined by the ratio η/d. Thus

$$L_n = f(d, \eta) . \tag{9.1}$$

The traditional approach of dimensional analysis gives (the dimensions of L_n, d and η are equal to L)

$$\Pi = \Phi(\Pi_1), \tag{9.2}$$

where $\Pi = L_n/d$ and $\Pi_1 = \eta/d$, so that

$$L_n = d\Phi\left(\frac{\eta}{d}\right) . \tag{9.3}$$

Now let the number of sides of the regular polygon approach infinity and the size length approach zero. From elementary geometry it is known that the perimeter of

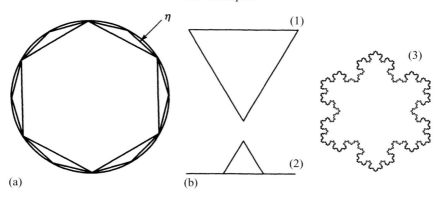

Figure 9.1 (a) A circle with inscribed regular polygons. As the number of sides in the polygon approaches infinity, and the side length approaches zero, the perimeter of the polygon approaches a finite limit. (b) A fractal curve, the Koch triad (von Koch, 1904): (1) the original triangle, (2) the elementary operation and (3) the broken line that approximates the fractal curve for a large number of sides. As the number of sides increases, the perimeter of the broken line approaches infinity according to a power law.

the inscribed regular polygon approaches a finite limit, $L_\infty = \pi d$, which is in fact the circumference of the circle. Thus, as $\eta/d \to 0$ and $n \to \infty$, the function Φ approaches a finite limit equal to π. Therefore at sufficiently large n (and small η) it is possible to replace the function Φ by its limit, equal to $\Phi(0, \infty) = \pi$, and to assume that the following relation is satisfied up to any desired accuracy for polygons with a large number of sides:

$$L_n = \pi d . \tag{9.4}$$

Consider now a curve, the *von Koch triad*, obtained in the following way (Figure 9.1b). An equilateral triangle of side d is taken, and each of its three sides is subjected to the following elementary operation: the side is divided into three equal sections and the middle section is replaced by two sides of an equilateral triangle constructed using it as the base. The sides of the polygon thus obtained are subjected to the same elementary operation, and so on. Obviously, the side length η_n of this polygon at the nth stage is equal to $d/3^n$, and the perimeter of the entire polygon L_n is equal to $3d(4/3)^n$. Equations (9.2) and (9.3) clearly hold in this case also. However, it can easily be shown that, since $n = \log(d/\eta)/\log 3$,

$$L_n = 3d[10^{n(\log 4 - \log 3)}] = 3d[10^{\alpha \log(d/\eta)}] = 3d\left(\frac{d}{\eta}\right)^\alpha , \tag{9.5}$$

where $\alpha = (\log 4 - \log 3)/\log 3 \cong 0.26\ldots$ So we find in this case the scaling law

$$\Phi\left(\frac{\eta}{d}\right) = 3\left(\frac{\eta}{d}\right)^{-\alpha}, \qquad L_n = 3d\left(\frac{\eta}{d}\right)^{-\alpha}, \qquad (9.6)$$

i.e. at $n = \infty$ the length of the curve L_n is infinite because only an empty relation, $\Pi = \infty$, is obtained in going to the limit $\eta/d \to 0$, $n \to \infty$.

However, usually one is interested not in the limit but in the asymptotics, in this case the perimeter of the polygon at large but finite n. It is not possible to pass to the limit and use a limiting relationship. However, relation (9.5) can be rewritten in the form

$$L_n = 3\frac{d^{1+\alpha}}{\eta^\alpha}. \qquad (9.7)$$

So a scaling, power-type, monomial relation is obtained for L_n but the structure of (9.7) cannot be determined by dimensional analysis because we do not know the number α beforehand. Furthermore, in this case the parameter η, and consequently the parameter n, remain in the resulting equation no matter how large n is, nor how small η is. Using as an approximation the relation $\Phi(\eta/d) \approx \Phi(0) = \text{const}$ and assuming, for large n, that $L_n \approx \text{const} \times d$ would lead to an incorrect result; in fact, at large n we find that L_n is proportional to $d^{1+\alpha}$, not to d.

The second fundamental example (which will be of special importance in Chapters 11 and 12, where turbulence is considered) is due to A. J. Chorin. This example elucidates a similar situation analytically. Consider a family of curves

$$\phi = \left(\ln\frac{d}{\delta}\right)\left(\frac{y}{\delta}\right)^{1/\ln(d/\delta)} - 2\ln\frac{d}{\delta}, \qquad (9.8)$$

where ϕ is a dimensionless function, d and δ are parameters with the dimension of length L and y is an independent variable also having the dimension of length. It is assumed that d is a fixed parameter, and that δ is a parameter of the family of curves.

It is easy to show that the family (9.8) satisfies the ordinary differential equation

$$\frac{d^2\phi}{dy^2} = \left(\frac{1}{\ln(d/\delta)} - 1\right)\frac{1}{y}\frac{d\phi}{dy}. \qquad (9.9)$$

A simple relation holds for the curves of the family

$$\left(y\frac{d\phi}{dy}\right) = \left(\frac{y}{\delta}\right)^{1/\ln(d/\delta)} = \exp\left(\frac{\ln(y/d) + \ln(d/\delta)}{\ln(d/\delta)}\right). \qquad (9.10)$$

This relation shows that as $d/\delta \to \infty$ and y/d remains fixed, the quantity $y(d\phi/dy)$ tends to a constant equal to e.

The family (9.8) having δ as a parameter has envelope

$$\phi = \ln \frac{y}{d} . \tag{9.11}$$

The quantity $y(d\phi/dy)$ for the envelope is a constant but a different one, equal to unity.

Assume now that in the differential equation (9.9) we neglect the term $1/\ln(d/\delta)$ in comparison with unity, which seems quite natural at large d/δ. Equation (9.9) reduces to the form

$$\frac{d^2\phi}{dy} = -\frac{1}{y}\frac{d\phi}{dy} . \tag{9.12}$$

Integrating this equation we obtain

$$\phi = C \ln y + D . \tag{9.13}$$

At $y = \delta$ the curves of the family (9.8) satisfy the boundary conditions

$$\phi(\delta) = -\ln\frac{d}{\delta}, \qquad \frac{d\phi}{dy}(\delta) = \frac{1}{\delta} . \tag{9.14}$$

The solution (9.13) satisfying the same boundary conditions as (9.14) comprises only one, d/δ-independent, curve, the envelope (9.11).

Now let us look at this matter from the viewpoint of classic dimensional analysis, as presented in Chapter 2. Following a standard procedure the derivative $d\phi/dy$ can be represented in the form

$$\frac{d\phi}{dy} = \frac{1}{y} \Phi\left(\frac{y}{\delta}, \frac{d}{\delta}\right) , \tag{9.15}$$

where Φ is a dimensionless function of its dimensionless arguments. Comparison with (9.8) gives the scaling law

$$\Phi = \left(\frac{y}{\delta}\right)^{1/\ln(d/\delta)} . \tag{9.16}$$

At arbitrary large y/δ and d/δ the function Φ cannot be replaced by a constant, so the influence of the parameter d/δ cannot be neglected.

These two examples clarify the situation for scaling laws which cannot be obtained by means of dimensional analysis.

9.2 Complete and incomplete similarity

Now we will look at this situation from a general viewpoint. In Chapter 2 we showed that any physically significant relation among, generally speaking, dimensional parameters,

$$a = f(a_1, \dots, a_k, b_1, b_2, b_3) , \tag{9.17}$$

can be represented in the form of a relation between dimensionless parameters,

$$\Pi = \Phi(\Pi_1, \Pi_2, \Pi_3), \qquad (9.18)$$

where

$$\Pi = \frac{a}{a_1^p \cdots a_k^r}, \quad \Pi_1 = \frac{b_1}{a_1^{p_1} \cdots a_k^{r_1}}, \quad \Pi_2 = \frac{b_2}{a_1^{p_2} \cdots a_k^{r_2}}, \quad \Pi_3 = \frac{b_3}{a_1^{p_3} \cdots a_k^{r_3}}. \qquad (9.19)$$

Here b_1, b_2, b_3 are the governing parameters, whose dimensions can be obtained as a product of powers of the dimensions of the parameters with independent dimensions $a_1, \ldots, a_k$. We have reduced the number of such parameters to three, because that is all we will need in this book. From (9.18) and (9.19) it follows that in such cases the relation of generalized homogeneity (see Chapter 2) of the function f looks like

$$f = a_1^p \cdots a_k^r \Phi_1 \left(\frac{b_1}{a_1^{p_1} \cdots a_k^{r_1}}, \frac{b_2}{a_1^{p_2} \cdots a_k^{r_2}}, \frac{b_3}{a_1^{p_3} \cdots a_k^{r_3}} \right). \qquad (9.20)$$

In traditional arguments "on a physical level" a parameter, say b_2, is considered to be substantial, i.e. actually to govern the phenomenon, if the value of the corresponding dimensionless parameter Π_2 is neither too large nor too small; more specifically, if its value is between $1/10$ and 10.

Thus, if a dimensionless governing parameter Π_2 corresponding to the dimensional parameter b_2 is small or large then, by a common tacit convention, it is assumed that its influence, and consequently that of the corresponding dimensional parameter b_2, can be neglected.

Actually, this argument is sometimes valid when there exists a finite non-zero limit of the function Φ in (9.18) as the parameter Π_2 goes to zero or infinity. In fact even more is required: the function Φ must converge sufficiently rapidly to its finite non-zero limit as Π_2 goes to zero or infinity. If these conditions are actually satisfied then, for sufficiently small or sufficiently large Π_2, the function Φ in (9.18) can be replaced by a function of two parameters,

$$\Pi = \Phi_1(\Pi_1, \Pi_3), \qquad (9.21)$$

where $\Phi_1(\Pi_1, \Pi_3) = \lim_{\Pi_2 \to 0, \infty} \Phi(\Pi_1, \Pi_2, \Pi_3)$.

We met such a situation in the problem of a very intense explosion, considered in Chapter 2. In fact in this problem, in addition to the governing parameters E, ρ_0, t, r and γ, there are two others, the initial radius of the shock wave r_0 and the pressure of the ambient quiescent air p_0. Of crucial importance in the arguments of G. I. Taylor and J. von Neumann was that they neglected those parameters when they were considering a concentrated and very intense explosion. Otherwise two additional dimensionless parameters would have appeared,

$$\frac{r_0}{\left(\dfrac{Et^2}{\rho_0}\right)^{1/5}} \ , \qquad \frac{p_0}{\left(\dfrac{\rho_0^{2/3} E^{2/5}}{t^{6/5}}\right)},$$

which would have made further analytical investigation impossible.

The reader may ask a natural question: in a real explosion, r_0 and p_0 are positive numbers which should definitely influence the whole gas motion from the very beginning to the very end. How can their values be taken to be equal to zero? That is, how is it possible to make the assumptions which Taylor and von Neumann made?

In fact, the real content of their assumptions was that they did not consider the whole motion from the very beginning to the very end but only the *intermediate stage*, when the radius of the shock wave r_f is much larger than the initial radius r_0 and, at the same time, the pressure behind the shock wave p_f remains much larger than the initial pressure p_0. At this stage the motion remains the same if r_0 is replaced by λr_0 and p_0 by μp_0, where λ and μ are arbitrary positive numbers of order unity (in fact this is the real content of their assumption).

In other words, the motion at this stage is invariant under an additional group of transformations, $r_0' = \lambda r_0$, $p_0' = \mu p_0$. There is no rigorous analytic proof of such invariance for this example, i.e. of the existence of a finite limit at $r_0 \to 0$, $p_0 \to 0$, but numerical computations and experimental results leave no doubt that this invariance holds.

In such cases we speak of *complete similarity, or similarity of the first kind, for a phenomenon in the parameter* Π_2. Another example of complete similarity is the first one considered above: the circle with inscribed regular *n*-gons (see Figure 9.1).

However, it is quite obvious that such a situation is far from being general. Usually, when a dimensionless parameter Π_2 goes to zero or infinity, the function $\Phi(\Pi_1, \Pi_2, \Pi_3)$ does not tend to any limit, let alone a finite and non-zero one. Therefore, in general, the parameter b_2 remains essential no matter how small or large is the value of the corresponding dimensionless parameter Π_2. This statement is correct, but trivial and non-constructive.

A fact of crucial importance is that *there exists a more general class of phenomena, which also is exceptional. However, this class is much wider than that of complete similarity. For phenomena of this class the function* Φ *entering* (9.18) *possesses at large or small values of* Π_2 *the property of generalized homogeneity in its own dimensionless arguments:*

$$\Phi = \Pi_2^\alpha \, \Phi\left(\frac{\Pi_1}{\Pi_2^\beta}, \Pi_3\right), \qquad (9.22)$$

where α and β are constants. This might seem to be exactly the same form of generalized homogeneity as for the basic function f in the relation (9.20). However, here there is a fundamental difference between the two cases. *The generalized homogeneity of the function f in* (9.20) *follows from the general physical invariance principle, and the constants* $p, \ldots, r_3$ *in* (9.20) *are obtained by simple rules of dimensional analysis. By contrast, the generalized homogeneity of the function in* (9.22) *is a special property of the problem under consideration.* In principle therefore, the constants α and β cannot be obtained using dimensional analysis; the relation (9.18) is the most that dimensional analysis can give.

In such exceptional cases the function f can be represented in the form

$$f = a_1^{p-\alpha p_2} \cdots a_k^{r-\alpha r_2} \, b_2^\alpha \, \Phi \left(\frac{b_1}{a_1^{p_1-\beta p_2} \cdots a_k^{r_1-\beta r_2} b_2^\beta} \,, \quad \frac{b_3}{a_1^{p_3} \cdots a_k^{r_3}} \right). \tag{9.23}$$

We will also meet an even more special case when $\beta = 0$; then we have

$$\Phi = \Pi_2^\alpha \, \Phi_1(\Pi_1, \Pi_3). \tag{9.24}$$

In this case

$$f = a_1^{p-\alpha p_2} \cdots a_k^{r-\alpha r_2} \, b_2^\alpha \, \Phi_1(\Pi_1, \Pi_3). \tag{9.25}$$

So, the general statement is this: the function f in these cases has the same property of generalized homogeneity as that contained in the Π-theorem (2.21), but the exponents cannot be obtained from dimensional analysis. Moreover, the parameter b_2 continues to influence the phenomenon. In such cases we speak of *incomplete similarity*, or *similarity of the second kind in the relevant parameter* Π_2.

The conclusion at which we have arrived is entirely natural: if the value of a certain governing dimensionless parameter Π_i is small or large then there are three possibilities.

(1) The limits of the corresponding functions Φ exist and are finite and non-zero as the dimensionless governing parameters Π_i tend to zero or infinity. The corresponding governing parameters, i.e. the dimensional b_i or the dimensionless Π_i, can be excluded from consideration and the number of arguments for the functions Φ therefore decreases. All similarity parameters can be determined by means of the regular procedure of dimensional analysis. This case corresponds to complete similarity of the phenomenon in the parameters Π_i.

(2) No finite limit exists for the function Φ as the Π_i tend to zero or infinity, but the special case indicated above holds. If so, the number of arguments of the function Φ can be decreased, but not all the parameters Π, Π_i can be obtained from dimensional analysis and the governing parameters b_i remain essential,

no matter how small or large are the corresponding similarity parameters. This case corresponds to incomplete similarity in the parameters Π_i.

(3) No finite limits exist for the function Φ as Π_i tend to zero or infinity, and the exceptions indicated in (1) and (2) do not hold. This case corresponds to a lack of similarity of the phenomenon in the parameters Π_i.

The difficulty is that, a priori, until we have obtained a mathematical solution of a complete non-idealized problem, we do not know with which of cases (1)–(3) we are dealing, irrespective of whether we have an explicit mathematical formulation of the problem or not. If such a solution is available, we do not need to use similarity methods. Hence we can only recommend assuming in succession each of the three possibilities for small or large similarity parameters – i.e. complete similarity, incomplete similarity or lack of similarity – and then compare the results obtained under each assumption with the data from experiments, numerical or physical, or the results of asymptotic analytical investigations. The term "experimental asymptotics", proposed for such analysis by Professor N. J. Zabusky, seems to be appropriate.

9.3 Self-similar solutions of the first and second kind

Consider a problem in mathematical physics that describes a certain phenomenon; let the quantity a be an unknown in this problem (it could be a "vector"), and let the quantities $a_1, \ldots, a_k, b_1, \ldots, b_n$ be independent variables and parameters appearing in the equations and in the boundary, initial and other conditions determining solutions.

Self-similar solutions are always the solutions of idealized (degenerate) problems that are obtained if certain parameters b_i, and the dimensionless parameters Π_i corresponding to them, assume zero or infinite values. They are simultaneously exact solutions of degenerate problems and asymptotic (generally speaking, intermediate asymptotic) representations of solutions of wider classes of non-idealized problems as the parameters b_i tend to zero or infinity.

It is clear that if an asymptotics is self-similar, and if the self-similar variables are power-law monomials, then one of the two special cases (1) and (2) mentioned above, i.e. that of complete or incomplete similarity, must hold. Correspondingly, self-similar solutions are divided into solutions of the first kind and second kind.

Self-similar solutions of the first kind. These are obtained when passage to the limit from a non-self-similar non-idealized problem to the corresponding self-similar idealized problem gives complete similarity in the parameters that made the original problem non-idealized and its solutions non-self-similar. Expressions for all the self-similar variables, independent as well as dependent, can be obtained by applying dimensional analysis.

Examples of such solutions, considered in previous chapters, are: the solutions found by G. I. Taylor and J. von Neumann for the problem of a very intense explosion; the Rayleigh problem of flow around a plane plate; and problems of a concentrated force or a concentrated force dipole in an elastic body.

Self-similar solutions of the second kind. Later we will consider several examples of incomplete similarity and self-similar solutions of the second kind. Here we simply mention that the Prandtl–Blasius solution for the problem of flow around a flat plane, considered in the boundary layer approximation in Chapter 8, also gives an example of a self-similar solution of the second kind. Indeed, this is an asymptotic solution valid for high values of the Reynolds number $Re = U\ell/\nu$ and large, but not too large, values of the parameter $x_1/(U/\nu)$. Dimensional analysis gives, we recall, for the velocity components u_1, u_2,

$$\Pi_{u1} = \frac{u_1}{U} = \Phi_1(\Pi_1, \Pi_2, Re) \,,$$

$$\Pi_{u2} = \frac{u_2}{U} = \Phi_2(\Pi_1, \Pi_2, Re) \,, \tag{9.26}$$

where $\Pi_1 = \xi_1 = Ux_1/\nu$ and $\Pi_2 = \xi_2 = Ux_2/\nu$. Here we have explicitly taken the Reynolds number into consideration. As was shown (formula (8.32)), at large values of Re the functions Φ_1 and Φ_2 take forms that are independent of the Reynolds number:

$$\Phi_1 = f_1\left(\frac{\Pi_2}{\sqrt{\Pi_1}}\right) \,, \qquad \Phi_2 = \frac{1}{\sqrt{\Pi_1}} f_2\left(\frac{\Pi_2}{\sqrt{\Pi_1}}\right) \,. \tag{9.27}$$

We recall that the asymptotic form of the solution (9.27) was obtained in Section 8.3 not by dimensional analysis only but using additionally the invariance of the asymptotics with respect to a transformation group. As we will see later, this is always the case.

9.4 Incomplete similarity in fatigue experiments (Paris' law)

It is worthwhile to return to Paris' law (6.23) for fatigue and consider it from the viewpoint presented above.

We assume that the shape of the loading cycle is fixed. Then the dependent quantity, the mean crack velocity averaged over the cycle $d\ell/dn$, can depend in principle upon the following arguments: the stress intensity-factor amplitude $\Delta N = N_{max} - N_{min}$, the stress-intensity-factor asymmetry $R = N_{max}/N_{min}$ (recall that N_{max} and N_{min} are the maximum and minimum values of the stress intensity factor over the cycle) and, what is especially important, the characteristic length scale h of the specimen, e.g. its diameter or thickness. We should also include in the list of governing parameters two important material properties: the yield stress σ_Y (analysis of the fracture surface shows that the local yield takes place in the upper part of the cycle) and a fracture toughness parameter. For the latter it is reasonable to take Irwin's parameter K_{Ic} and not the cohesion modulus K, because the crack extension goes by jumps, i.e. unstably. Thus, we assume that there exists a relation of the form

$$\frac{d\ell}{dn} = f(\Delta N, R, \sigma_Y, K_{Ic}, h) . \tag{9.28}$$

The dimensions of the parameters entering (9.28) are as follows:

$$[N] = [K_{Ic}] = FL^{-3/2}, \qquad [\sigma_Y] = FL^{-2}, \qquad [h] = L, \qquad [R] = [n] = 1 . \tag{9.29}$$

Here F is the dimension of force and L is the dimension of length. The dimension of the quantity $d\ell/dn$ is therefore equal to L.

We take as the governing parameters, with independent dimensions, the parameters ΔN and σ_Y. Dimensional analysis gives by the standard procedure

$$\frac{d\ell}{dn} = \left(\frac{\Delta N}{\sigma_Y}\right)^2 \Phi\left(\frac{\Delta N}{K_{Ic}}, R, Z\right) , \tag{9.30}$$

where the dimensionless parameter

$$Z = \frac{\sigma_Y \sqrt{h}}{K_{Ic}} \tag{9.31}$$

is the square root of the ratio of the characteristic specimen length scale h and the fracture yield scale σ_γ^2/K_{Ic}^2.

Estimates show that the dimensionless parameter $\Pi_1 = \Delta N/K_{Ic}$ is small. Thus here we have an appropriate case for applying the technique of analysis of asymptotic scaling laws presented above. Thus, if a finite non-zero limit of the function Φ in (9.30) as $\Delta N/K_{Ic} \to 0$ did exist, i.e. if there were *complete similarity* in the parameter $\Delta N/K_{Ic}$, we would obtain the scaling law

$$\frac{d\ell}{dn} = \left(\frac{\Delta N}{\sigma_Y}\right)^2 \Phi_1(R, Z), \tag{9.32}$$

so that the parameter m in the Paris scaling law (6.23) would equal 2; however, in (9.32) the form of the constant A from (6.23) appears to be non-universal, since it will depend on the asymmetry of the loading cycle and the specimen size. Analysis of the experimental data shows that $m = 2$ is practically never found: for some aluminium alloys m is close to 2, but nevertheless it is always larger than, 2. For the vast majority of cases m is substantially larger than 2.

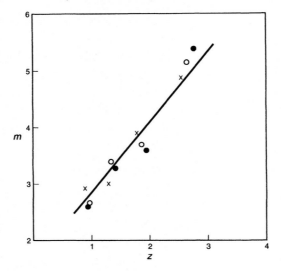

Figure 9.2 The dependence of the exponent m in Paris' law on the similarity parameter Z for 4340 for the specimens of various orientation (A, L, T) with respect to rolling direction: A (hollow circles), L (full circles), T (crosses). From Barenblatt and Botvina (1981).

Therefore, assuming incomplete similarity we obtain (Barenblatt and Botvina, 1981; Carpinteri, 1996)

$$\Phi = \left(\frac{\Delta N}{K_{\mathrm{Ic}}}\right)^{\alpha(R,Z)} \Phi_1(R, Z) \,, \tag{9.33}$$

which is exactly the form of the experimentally observed Paris law (6.23), with the following expressions for the parameters of this law:

$$A = \frac{\Phi_1(R, Z)}{\sigma_Y^2 K_{\mathrm{Ic}}^{\alpha}} \,, \qquad m = 2 + \alpha(R, Z) \,. \tag{9.34}$$

The most important conclusion of the analysis just performed is that the parameters A and m of Paris' law *are not material characteristics*. They must depend upon, besides the asymmetry of the cycle R, the specimen's length scale.

This conclusion is of great significance, and it had to be checked experimentally. Indeed, persuasive experimental data obtained by Botvina (who analyzed the data of Heiser and Mortimer; see Barenblatt and Botvina (1981)) and by Ritchie (2005) (with data from Ritchie and Knott (1974); see Figures 9.2 and 9.3) showed that the dependence of m upon Z, i.e. the specimen size, can be substantial. Therefore, using the results of standard fatigue experiments performed on small specimens in practical structural design can be dangerous: the real lifetime of a structure can be overestimated.

This example has a wider meaning. Power laws are often used in engineering practice as material properties. Characteristic examples are calculations for

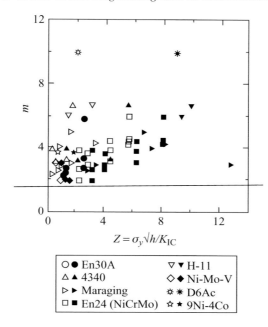

Figure 9.3 The dependence of the exponent m in Paris' law on the similarity parameter Z for various steels. From Ritchie, 2005 with kind permission from Springer Science+Business Media.

plastic materials assuming a power-law constitutive equation and the evaluation of the lifetime of polymeric structures assuming a power-law dependence of the fracture toughness on the crack-tip velocity. *In fact, the universality of the constitutive relations should be checked carefully for specimens of various sizes, otherwise predictions of strength could be unreliable.* This is one of many results obtained in recent decades and it is based on the concept of incomplete similarity; see Carpinteri (1996) and Bažant (2002). There are many other problems, in particular, in turbulence, where this approach has led to a reconsideration of seemingly well-established views. See more about these results in Chapters 11 and 12.

9.5 A note concerning scaling laws in nanomechanics

Micromechanics, by definition, considers phenomena of characteristic length scale $1\,\mu m = 10^{-4}$ cm, with a "pixel" size, i.e. internal length scale, of the order of 10^{-6}–10^{-5} cm. This scale is never included in the set of governing parameters because it is selected rather arbitrarily.

The situation is different for nanomechanics, where the pixel size, $\lambda \sim 10^{-10}$ m $= 10^{-8}$ cm, has a physical meaning and therefore must be included in the set of governing parameters. We restrict ourselves here to deformation and strength phenomena. Clearly these phenomena are influenced by a microscopic Young modulus E, which also determines the maximum cohesion force. Furthermore, the density ρ of the material is of substantial importance for phenomena in nanomechanics.

Indeed, even in macroscopically steady processes, such as steady deformation, the multiple nanoscale defects present in the material interact with each other. This interaction can lead to dynamic effects, similar to the breaking of a crack through the couple of stiffeners considered in Chapter 6: the role of the stiffeners is played by neighboring defects. These dynamic effects are influenced by the density. Furthermore, when we are considering a nanomechanical range of scale, quantum-mechanical effects can be expected; therefore the Planck constant h should enter the set of governing parameters. The dimensions of all these parameters are:

$$[E] = \frac{M}{LT^2} , \qquad [\rho] = ML^{-3} , \qquad [h] = \frac{ML^2}{T} .$$

Here L, M and T are, as usual, the dimensions of length, mass and time.

What is critical here is that from the parameters E, ρ and h a parameter with the dimension of length can be formed:

$$\lambda = \left(\frac{h}{(\rho E)^{1/2}} \right)^{1/4} = \left(\frac{h}{\rho \sqrt{E/\rho}} \right)^{1/4} . \qquad (9.35)$$

Note immediately the small exponent $\frac{1}{4}$. Now let us take the actual values of the parameters. The value of the Planck constant h is $\sim 7 \times 10^{-34}$ m^2 kg/s. The density varies from that for platinum ($\rho = 21.0 \times 10^3$ kg/m^3), tungsten and plutonium ($\rho = 19.0 \times 10^3$ kg/m^3), and so on, to that for water ($\rho = 1 \times 10^3$ kg/m^3). The quantity $(E/\rho)^{1/2}$ is equal to a constant times 10^3 m/s, where the constant is of the order of 1 (namely 1.5 for water, 5.0 for rock etc.). Substituting into (9.35) we obtain

$$\lambda = \left(\frac{7 \times 10^{-34}}{B \times 10^6} \right)^{1/4} = A(10^{-40})^{1/4} \approx 10^{-10} \text{ m} = 10^{-8} \text{ cm} .$$

Here B is a constant between 1 and 100, and the constant A is always of the order of 1 (owing to the smallness of the exponent $\frac{1}{4}$).

We notice that $\lambda = 10^{-10}$ m $= 10^{-8}$ cm is exactly the size of the pixel in nano-phenomena and so, in contrast with the size of the pixel in micromechanics, λ is a fundamental physical quantity. In particular the normal interatomic distance in a crystal lattice, 1 ångström, is of this order.

Obviously this fact can have important physical consequences: the parameter λ should enter the set of governing parameters. Therefore the dimensionless parameter Λ/λ, where Λ is the length scale of a structure or specimen, should enter the set of governing dimensionless parameters.

This means that, in cases of incomplete similarity or even a lack of similarity in Λ/λ, scaling effects should be observed at large Λ/λ. In particular, large macroscopic structures or specimens cannot be excluded, and so this is a plausible way of observing nanomechanical (quantum-mechanical) effects on fatigue, fracture, plasticity and other mechanical properties of large specimens and structures (Barenblatt and Monteiro, 2010).

10

The ideal gas approximation. Sound waves; shock waves

The model to be presented in this chapter is based on the assumption that the fluid (gas) is ideal but not incompressible. This model has had in the past, and continues to have, many practical applications of the first importance, particularly in aviation problems. It originated in the middle of the nineteenth century, continued into the beginning of the last century, and was studied most intensively from the 1930s to the 1950s. However, research on this model is far from complete. Many dark areas of fundamental and practical interest remain. To be specific I will mention one of them.

At the wing of an aircraft moving with a large subsonic velocity a "local supersonic zone" is formed. This zone is bounded at the front by a "sonic surface" where the velocity is equal to the velocity of sound and at the rear by a shock wave; this is a well-established experimental fact. The structure of the intersection of the sonic surface and the shock wave is not yet known in spite of long-time efforts by first-class specialists over several decades.

There are nowadays many textbooks, treatises and monographs in which the current state of the ideal gas model is presented in detail; I give special mention to Oswatisch (1956), Zeldovich and Raizer (2002) and Hayes and Probstein (2004). In this chapter we will consider some instructive topics which, as a rule, are not covered in textbooks.

10.1 Sound waves

Sound waves give an instructive application of the model under consideration. To begin we recall the fundamental thermodynamic relations. The first law of thermodynamics can be expressed in the form

$$\delta Q = d\mathcal{E} + \delta A \,, \tag{10.1}$$

where δQ is the elementary inflow of heat into a body, $d\mathcal{E}$ is the total differential of the *internal energy* $\mathcal{E}$, which is a function of the parameters of state, and δA is the elementary work performed by the body against the ambient medium. For an ideal gas, $\delta A = p\,d(1/\rho)$ per unit mass. It is important that δQ is not a total differential, but expression (10.1) has an integrating factor $1/T$, where T is the Kelvin absolute temperature. Therefore $\delta Q/T = dS$, where S, the *entropy*, is also a function of the parameters of state. The equation of state of a thermodynamically perfect gas will be used further:

$$p = R\rho T \tag{10.2}$$

where R is the gas constant per unit mass of gas. The specific (per unit mass) internal energy for a thermodynamically perfect gas is

$$\mathcal{E} = c_v T \;. \tag{10.3}$$

Substituting (10.3) into (10.1), we obtain that c_v is the specific heat of such a gas at constant volume (density). The relation (10.1) can be represented also in the form

$$\delta Q = T\,dS = d\left(\mathcal{E} + \frac{p}{\rho}\right) - \frac{1}{\rho}\,dp \;. \tag{10.4}$$

The function $W = \mathcal{E} + p/\rho = (c_v + R)T$, called the *enthalpy*, or heat function, is also a function of the parameters of state. So, δQ can be represented in the form $\delta Q = c_p\,dT - \frac{1}{\rho}\,dp$, where $c_p = c_v + R$ is the specific heat at constant pressure. Combining the relations (10.1)–(10.4) we obtain an expression for the entropy per unit mass:

$$S = c_v \ln \frac{p}{\rho^\gamma} \;. \tag{10.5}$$

Here the exponent $\gamma = c_p/c_v$ is called the adiabatic index.

Now we consider a sound wave of frequency ω and wavelength λ. The phenomenon of sound propagation is characterized by the smallness of the pressure and density perturbations:

$$p = p_0 + p', \qquad \rho = \rho_0 + \rho', \qquad \left|\frac{p'}{p_0}\right| \ll 1 \;, \qquad \left|\frac{\rho'}{\rho_0}\right| \ll 1 \;. \tag{10.6}$$

Substituting (10.6) into the continuity equation,

$$\partial_t \rho + \nabla \rho \cdot \mathbf{u} = 0 \;, \tag{10.7}$$

we obtain as a first approximation

$$\partial_t (\rho'/\rho_0) + \nabla \cdot \mathbf{u} = 0 \;. \tag{10.8}$$

In contrast with the case of an ideal incompressible fluid (see Chapter 3), the terms in the divergence of the velocity vector $\partial_\alpha u_\alpha$ that are of order u_α/λ have the same magnitude as the term $\partial_t(\rho'/\rho_0)$. The latter has magnitude $(\rho'/\rho_0)\omega$;

therefore the velocity components are of order $(\lambda\omega)(\rho'/\rho)$, i.e. they are first-order quantities. Thus in the Euler equation of momentum balance,

$$\partial_t \mathbf{u} + (\mathbf{u} \cdot \nabla) \cdot \mathbf{u} = -\frac{1}{\rho}\nabla p \, , \qquad (10.9)$$

the nonlinear term can be neglected and so the momentum equation takes the form

$$\partial_t \mathbf{u} = -\frac{1}{\rho_0}\nabla p' \, . \qquad (10.10)$$

Furthermore, the propagation of sound waves is fast, so that the heat exchange of the fluid particles can be neglected and the entropy is preserved ($\delta Q = TdS = 0$). Assuming that the medium is homogeneous, we obtain in the approximation (10.6)

$$p' = \left(\frac{\partial p}{\partial\rho}\right)_{S=\text{const}} \rho' = \frac{\gamma p_0}{\rho_0}\rho' \, . \qquad (10.11)$$

Collecting together equations (10.8), (10.10) and (10.11), we obtain the basic equation for the pressure perturbation, the *D'Alembert equation*

$$\partial_{tt}^2 p' = c^2 \Delta p' \, , \qquad (10.12)$$

where $c^2 = (\gamma p_0/\rho_0)$.

In the one-dimensional case the D'Alembert equation takes the form

$$\partial_{tt}^2 p' = c^2 \partial_{xx}^2 p', \qquad (10.13)$$

where x is the space coordinate. The general solution to this equation is a sum of two "traveling waves", propagating in opposite directions without changing their form:

$$p' = F(x + ct) + G(x - ct) \, . \qquad (10.14)$$

The sound velocity c is a fundamental characteristic of the fluid state. In particular, there is a principal difference in the propagation of small perturbations from a moving body in a gas depending on whether the body speed u is subsonic ($u < c$) or supersonic ($u > c$). If the body speed is larger than the velocity of sound, $u > c$, the perturbations do not penetrate the whole space; they are concentrated (Figure 10.1) inside a cone, called the *Mach cone*. The angle 2α at the Mach cone vertex satisfies the relation $\sin\alpha = c/u = 1/M$, where M is a similarity parameter called the *Mach number* (see Chapter 2). If $u < c$ then the perturbations propagate ahead of the moving body as well.

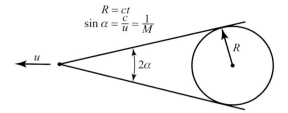

Figure 10.1 The Mach cone. The perturbations are concentrated within a cone. The angle 2α at the cone vertex satisfies the relation $\sin \alpha = c/u = 1/M$.

10.2 Energy equation. The basic equations of the ideal gas model

To obtain a closed system of equations for the model we add to the equations of ideal fluid the continuity equation obtained from the mass conservation law,

$$\partial_t \rho + \nabla \cdot \rho \mathbf{u} = 0 \qquad (\partial_t \rho + \partial_\alpha \rho u_\alpha = 0) \tag{10.15}$$

and the Euler equation obtained from the momentum conservation law,

$$\partial_t \rho \mathbf{u} + \nabla \cdot (\rho \mathbf{u} \otimes \mathbf{u}) + \nabla p = 0 \qquad (\partial_t \rho u_i + \partial_\alpha \rho u_i u_\alpha + \partial_i p = 0) , \tag{10.16}$$

an energy balance equation. In the case of an ideal, thermodynamically perfect, gas these three equations form a closed description of the model.

As in previous cases we consider a fixed region Ω bound by a surface $\partial\Omega$. The time derivative of the total energy of the body occupying the region Ω as its configuration at time t is composed of the flow of total energy together with the fluid flow through the boundary $\partial\Omega$ and the work performed by the body against the pressure of the ambient fluid:

$$\partial_t \int_\Omega \left(\rho \frac{\mathbf{u}^2}{2} + \rho \mathcal{E} \right) d\omega = - \int_{\partial\Omega} \rho \left(\frac{\mathbf{u}^2}{2} + \mathcal{E} \right) \mathbf{u} \cdot \mathbf{n} \, d\Sigma - \int_{\partial\Omega} p \mathbf{n} \cdot \mathbf{u} \, d\Sigma$$

$$= - \int_{\partial\Omega} \rho \left(\frac{\mathbf{u}^2}{2} + \mathcal{E} + \frac{p}{\rho} \right) \mathbf{u} \cdot \mathbf{n} \, d\Sigma , \tag{10.17}$$

so that the rate of variation of the total energy of a body having Ω as its configuration is balanced by the flux of the kinetic energy and enthalpy. We do not take into account this time the energy inflow into the volume Ω due for example to a chemical reaction. In a standard way demonstrated above for the mass and momentum balance equations we obtain a differential equation for the energy balance:

$$\partial_t \left(\rho \frac{\mathbf{u}^2}{2} + \rho \mathcal{E} \right) + \partial_\alpha \left[\rho u_\alpha \left(\frac{\mathbf{u}^2}{2} + \mathcal{E} + \frac{p}{\rho} \right) \right] = 0 . \tag{10.18}$$

Using the equations (10.2), (10.3), (10.5), (10.15) and (10.16) we reduce equation (10.18) to an equation of entropy conservation for a fluid particle:

$$\frac{d}{dt}\left(\frac{p}{\rho^\gamma}\right) = \partial_t\left(\frac{p}{\rho^\gamma}\right) + u_\alpha\partial_\alpha\left(\frac{p}{\rho^\gamma}\right) = 0 . \tag{10.19}$$

The system (10.15), (10.16), (10.19) forms a closed system of equations for the ideal gas model. Appropriate initial and boundary conditions are needed to complement these equations. We emphasize that in contrast with the ideal incompressible fluid model an initial condition for the pressure should be prescribed since equation (10.19) contains the time derivative of pressure.

10.3 Simple waves. The formation of shock waves

Compressibility can lead to a new phenomenon, non-existent for the model from incompressible fluid: *shock waves*, thin regions of sharp variations in the flow properties. To observe the formation of shocks we consider a classical solution to the basic equations of *isentropic flow* (10.15), (10.16); it is assumed that at the initial moment the entropy is constant in the entire space so that, owing to equation (10.19), it remains constant:

$$p = C\rho^\gamma . \tag{10.20}$$

The solution, that we present here, obtained by the great German mathematician B. Riemann, describes one-dimensional motion for which the velocity u is a *single-valued* function of density: $u = u(\rho)$. All the flow properties depend on the space coordinate and time t. Using equations (10.15) and (10.16) we obtain

$$\frac{du}{d\rho}\partial_t\rho + \left(u(\rho)\frac{du}{d\rho} + \frac{1}{\rho}\frac{dp}{d\rho}\right)\partial_x\rho = 0 , \tag{10.21}$$

$$\partial_t\rho + \left(\rho\frac{du}{d\rho} + u\right)\partial_x\rho = 0 . \tag{10.22}$$

From these equations, bearing in mind that $c^2 = dp/d\rho$, it follows that

$$\frac{du}{d\rho} = \pm\frac{c}{\rho} , \qquad u = \pm\int\frac{dp}{\rho c} . \tag{10.23}$$

This relation concludes the solution.

Now consider the speed of propagation of a point in space (not a particle!) $x = x_c(t)$ at which the density (and also the velocity and pressure) has a fixed value: $\rho = \rho(x_c(t), t)$ is a constant. We have

$$(\partial_x\rho)\frac{dx_c}{dt} + \partial_t\rho = 0 , \qquad \frac{dx_c}{dt} = -\frac{\partial_t\rho}{\partial_x\rho} , \tag{10.24}$$

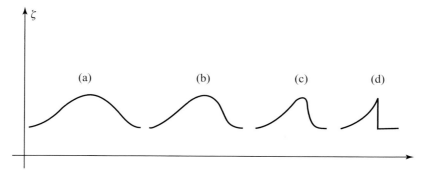

Figure 10.2 Evolution of the density distribution in a simple wave. (a) The initial distribution, which has the shape of a hump; (b) the slope of leading edge of the hump becomes steeper; (c) the density (velocity, pressure) gradient becomes so large that the ideal fluid approximation becomes invalid in a thin zone; (d) the density (velocity, pressure) distribution after the shock wave has appeared.

and so, using the continuity equation (10.15), we obtain

$$\frac{\partial_t \rho}{\partial_x \rho} = -\left(u + \rho \frac{du}{d\rho}\right) , \qquad \frac{dx_c}{dt} = u + \rho \frac{du}{d\rho} = u \pm c(u) . \qquad (10.25)$$

Thus, the speed of propagation of the constant value of density (and also of pressure and velocity) is composed of the fluid velocity at the point *and the local velocity of sound.* Compare this with a sound wave, for which the gas velocity u is small and the local sound velocity is equal to a constant, c_0, so that the sound wave propagates as a traveling wave of fixed shape.

In contrast with a sound wave, points in space where the density (and, according to (10.20), (10.23), the pressure and velocity) is larger propagate faster. Thus the evolution of the density distribution, which at the beginning has the shape of a hump, is distorted (see Figure 10.2). Most importantly, at a certain moment the density (velocity, pressure) gradient becomes so large that the ideal fluid approximation becomes invalid: the dissipative processes related to the gradients become intensified in a narrow region at the hump's leading edge. However, consideration of the flow can be continued within the framework of the same model but using "weak solutions" that admit discontinuities (these are the "generalized solutions" in the terminology of S. L. Sobolev, who in the mid 1930s pioneered the consideration of such solutions of partial differential equations). These discontinuities model shock waves, a well-known phenomenon in gas dynamics (see e.g. Figure 2.1). Naturally, certain conditions are needed for matching the flow features behind and ahead of the discontinuities.

To derive these conditions we return to the integral equations for the balance of mass (1.12), momentum (1.24) and energy (10.17). All these equations have a

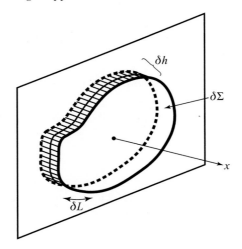

Figure 10.3 Derivation of the conditions at the shock wave; see the main text.

similar form:

$$\partial_t \int_\Omega A \, d\omega + \int_{\partial\Omega} \mathbf{B} \cdot \mathbf{n} \, d\Sigma = 0 \,, \tag{10.26}$$

where $A = \rho$, $\mathbf{B} = \rho\mathbf{u}$ for the mass balance, $\mathbf{A} = \rho\mathbf{u}$, $\mathbf{B} = \rho\mathbf{u} \otimes \mathbf{u} + p\mathbf{I}$ for the momentum balance and $A = \rho\mathbf{u}^2/2 + \mathcal{E}$, $\mathbf{B} = \rho(\mathbf{u}^2/2 + \mathcal{E} + p/\rho)\mathbf{u}$ for the energy balance.

To derive the conditions at the shock fronts we choose as Ω a thin region (see Figure 10.3) moving with the shock front velocity D in the direction normal to the front. (For the sake of simplification we consider the one-dimensional case, which is all that we will need later.) The region Ω is bound by two elementary areas parallel to the shock front (one in front of it and the other behind it) and having area $\delta\Sigma$, and a cylindrical surface, normal to the front, having height δh and contour length δL. Now we pass to the limit $\delta h \to 0$, leaving $\delta\Sigma$ intact. The integral over the volume, which is of order $\delta h \delta\Sigma$, vanishes as does the integral over the surface of the cylinder, which is of order $\delta h \delta L$. What remains is a term proportional to $\delta\Sigma$,

$$\delta\Sigma(B_{n+} - B_{n-}) \,, \tag{10.27}$$

so that the conditions of matching the flow properties behind and ahead of the front – i.e. the surface of discontinuity – take the general form $B_{n+} - B_{n-} = 0$ or, separately, for the mass,

$$\rho_+(u_+ - D) = \rho_-(u_- - D); \tag{10.28}$$

for the momentum,

$$\rho_+(u_+ - D)^2 + p_+ = \rho_-(u_- - D)^2 + p_- ; \tag{10.29}$$

and, for the energy,

$$\rho_+(u_+ - D)\left[\frac{(u_+ - D)^2}{2} + \frac{\gamma}{g-1}\frac{p_+}{\rho_+}\right] = \rho_-(u_- - D)\left[\frac{(u_- - D)^2}{2} + \frac{\gamma}{\gamma-1}\frac{p_-}{\rho_-}\right].$$
$$\tag{10.30}$$

Here the subscript plus refers to values ahead of the shock front and the subscript minus refers to values behind the shock front. The conditions (10.28)–(10.30) for matching at the discontinuity surface – the shock front – conclude the model.

10.4 An intense explosion at a plane interface: the external intermediate asymptotics

The problem of an intense explosion at the interface of two half-spaces each containing the same gas but with different densities and separated initially by an impermeable wall is an instructive example of flows with shock waves and also of multiscale intermediate asymptotics and self-similar solutions of both the first and second kind.

Let the gas in the right-hand half-space ($x \geq 0$) have density ρ_0 and that in the left-hand half-space ($x \leq 0$) have a density ρ_1 that is much less than ρ_0: $\rho_1 \ll \rho_0$. At the moment $t = 0$ a large amount of energy is generated in a small vicinity of the wall inside the right-hand half-space. We assume at first that the energy generation is instantaneous and concentrated at $x = +0$. Immediately after the explosion the wall is instantaneously removed, and gas starts to move both to the right and to the left, so that a shock wave propagates in each half-space. An intermediate stage at which both shock waves are very intensive, so that the pressures behind them are much larger than those ahead, will be considered.

The solution to this problem was constructed by R. I. Nigmatulin (1965); the solution for a symmetric planar explosion obtained by L. I. Sedov (1946, 1959) was used in this construction.

The conditions at the very intense right-hand shock wave, which propagates into the half-space containing gas of higher density, can be reduced to the form

$$\rho_{f2}(u_{f2} - D_2) = -\rho_0 D_2 , \tag{10.31}$$

$$\rho_{f2}(u_{f2} - D_2)^2 + p_{f2} = -\rho_0 D_2^2 , \tag{10.32}$$

$$\rho_{f2}(u_{f2} - D_2)\left[\frac{\gamma p_{f2}}{(\gamma - 1)\rho_{f2}} + \frac{(u_{f2} - D_2)^2}{2}\right] = -\rho_0 \frac{D_2^3}{2} , \tag{10.33}$$

where D_2 is the velocity of propagation of the right-hand shock wave and the subscript 2 corresponds to quantities behind this wave. Analogous conditions hold for the left-hand shock wave, propagating in the negative direction with velocity D_1; the subscript 1 corresponds to the left-hand shock wave. The dimension of the energy E (per unit area) is MT^{-2}. The solution to this problem is self-similar; this can be shown by following for the case of plane waves the arguments of dimensional analysis exactly as used for spherical waves by G. I. Taylor and J. von Neumann (see Chapter 2).

The solution can be presented in the form

$$u = u_{f2} f(\zeta, \gamma), \qquad \rho = \rho_{f2} g(\zeta, \gamma), \qquad p = p_{f2} h(\zeta, \gamma). \qquad (10.34)$$

Here $\zeta = x/x_{f2}$; u_{f2}, ρ_{f2} and p_{f2} are the velocity, density and pressure immediately behind the right-hand front, which is moving according to the law $x = x_{f2}$ into the high-density region. Using dimensional analysis we also obtain

$$x_{f2} = \xi_0 \left(\frac{Et^2}{\rho_0} \right)^{1/3}, \qquad (10.35)$$

where ξ_0 is a constant that depends on γ only. For the left-hand shock wave, propagating into the low-density-gas half-space, using once again dimensional analysis we obtain

$$x_{f1} = \zeta_1 x_{f2}, \qquad (10.36)$$

so that the left-hand shock wave corresponds to a constant value $\zeta = \zeta_1 < 0$. The conditions at the right-hand shock wave and analogous conditions at the left-hand shock wave follow the relations

$$p_{f2} = \frac{2}{\gamma + 1} \rho_0 D_2^2, \qquad p_{f1} = \frac{2}{\gamma + 1} \rho_1 D_1^2. \qquad (10.37)$$

For the velocities of the shock waves D_1, D_2 the following relations are obtained:

$$D_1 = \tfrac{2}{3} \zeta_1 \xi_0 \left(\frac{E}{\rho_0} \right)^{1/3} \frac{1}{t^{1/3}} = \zeta_1 D_2,$$

$$D_2 = \tfrac{2}{3} \xi_0 \left(\frac{E}{\rho_0} \right)^{1/3} \frac{1}{t^{1/3}}. \qquad (10.38)$$

In fact the motion established in each of the half-spaces $x \lessgtr 0$ corresponds to a symmetric very intense explosion with energies E_1, E_2 and densities ρ_1, ρ_0 respectively. We note that for a symmetric explosion the pressure $p(0, t)$ at the plane $x = 0$ is a fixed, time-independent, part of the pressure at the front, depending only on the adiabatic index γ. Therefore, from the condition of the pressure continuity at $x = 0$

it follows that $p_{f1} = p_{f2}$ and, using the relations (10.35)–(10.38), we obtain

$$\zeta_1 = -\sqrt{\frac{\rho_0}{\rho_1}} \,, \qquad \frac{E_2}{E_1} = \sqrt{\frac{\rho_1}{\rho_0}} \,. \tag{10.39}$$

In particular, it follows from (10.39) that in the limit $\rho_1/\rho_0 \to 0$ the whole energy goes instantaneously into a vacuum.

10.5 An intense explosion at a plane interface: the internal intermediate asymptotics

Nigmatulin's intermediate asymptotic solution, considered in the previous section, is valid in the time interval $\mathcal{T} \gg t \gg \tau$, where $\mathcal{T}$ is the time when the pressure at the right-hand shock front becomes comparable with the initial pressure in the right-hand half-space p_0,

$$\mathcal{T} \sim \frac{E_2 \rho_0^{1/2}}{p_0^{3/2}} \sim \frac{E_1 \rho_1^{1/2}}{p_0^{3/2}} \,,$$

and τ is the duration of energy release, which in fact is also finite; the effects of a concentrated explosion do not settle down immediately.

To clarify the situation we will consider the initial conditions in more detail. We assume that the pressure at the right-hand side of the impermeable wall increases sharply but not instantaneously, according to the law $p(0+, t) = p_{\mathrm{w}} f(t/\tau)$, up to a certain finite time $t = \tau$, and only after that does the impermeable wall disappear instantaneously. Here $p_{\mathrm{w}} \gg p_0$ and τ are constants and $f(t/\tau)$ is a certain function, of the order of 1, that is positive when its argument is less than 1, equal to 1 at $t = 0$ and equal to 0 at $t/\tau \geq 1$. The subsequent flow develops as follows.

A shock wave $x = x_f(t/\tau)$ propagates to the right in the quiescent gas. In a certain region behind the shock wave the compressed gas continues to advance to the right. At a certain plane $x = x_i(t)$, however, the instantaneous speed of the gas particles becomes equal to zero and all gas particles situated to the left of this plane move to the left: there occurs an expansion into the low-density region of gas, which is thus compressed by the shock wave.

This problem was formulated and solved independently by von Weizsäcker (1954), and Zeldovich (1956). The problem was reduced to the solution of a system consisting of the differential equations of mass and momentum balance

$$\partial_t \rho + \partial_x(\rho u) = 0 \,,$$
$$\partial_t u + u \partial_x u + \frac{1}{\rho} \partial_x p = 0 \,, \tag{10.40}$$

complemented by the differential equation of conservation of entropy

$$\partial_t \left(\frac{p}{\rho^\gamma} \right) + u \partial_x \left(\frac{p}{\rho^\gamma} \right) = 0 \,.$$

The last differential equation is necessary because the flow is not isentropic: the entropy jump at the shock front is time dependent.

The boundary conditions (10.26)–(10.28) at the shock wave $x = x_f(t)$ moving with speed D_2 take the form

$$\rho_f(u_f - D_2) = -\rho_0 D_2 \,,$$
$$\rho_f(u_f - D_2)u_f + p_f = 0 \,,$$
$$\rho_f(u_f - D_2)\left[\frac{u_f^2}{2} + \frac{p_f}{(\gamma - 1)\rho_f} \right] + p_f u_f = 0 \,.$$
(10.41)

Here again we have taken into account that the pressure p_f behind the shock wave is much larger than initial pressure p_0.

The initial conditions at time $t = \tau$ correspond for $x < 0$ to the gas being at rest under zero pressure and density ρ_1 and for $x > 0$ to the state of motion that has developed at time $t = \tau$, in the half-space filled with quiescent gas of density ρ_0 at zero pressure, owing to the maintenance on the boundary during the time interval τ of a pressure varying according to the law $p(w, t) = p_2 f(t/\tau)$.

It is evident that the density, pressure and speed of the gas depend on the dimensional quantities

$$t, \ p_w, \ \rho_0, \ \tau, \ x \tag{10.42}$$

and that the coordinate of the shock front $x_f(t)$ depends on all these quantities except the last. Applying the standard procedure of dimensional analysis, we obtain

$$x_f(t) = \sqrt{p_w/\rho_0} \ \tau \xi_f(\Pi_1) \,, \tag{10.43}$$

$$\Pi_\rho = \Phi_\rho(\Pi_1, \Pi_2) \,, \qquad \Pi_p = \Phi_p(\Pi_1, \Pi_2) \,, \qquad \Pi_u = \Phi_u(\Pi_1, \Pi_2) \,. \tag{10.44}$$

Here

$$\Pi_1 = \frac{t}{\tau} = \theta \,, \qquad \Pi_2 = \frac{x}{\sqrt{p_w/\rho_0}\tau} \,,$$

$$\Pi_\rho = \frac{\rho}{\rho_0} \,, \qquad \Pi_p = \frac{p}{p_w} \,, \qquad \Pi_u = \frac{u}{\sqrt{p_w/\rho_0}} \,. \tag{10.45}$$

As is evident, the solution to the problem posed turns out to be non-self-similar. This results from the fact that the problem contains a characteristic time τ and a characteristic length scale $(p_w/\rho_0)^{1/2}\tau$. Numerical calculations reveal, however, that the solution to the formulated problem has an instructive property. Namely, the dependence of the coordinate of the wavefront on time rapidly (i.e. after a time interval of order τ) approaches a scaling, power-law, asymptotics, so that

$$\xi_f(\Pi_1) = \xi_0(\gamma)\Pi_1^\alpha , \qquad (10.46)$$

where $\xi_0(\gamma)$ is some function of γ and the exponent α also depends on γ. Furthermore, the density at the front rapidly approaches a constant value, and the pressure and speed of the gas at the front rapidly approach the scaling laws

$$\frac{p_f}{p_w} \sim \Pi_1^{-2(1-\alpha)} , \qquad \frac{u_f}{\sqrt{p_w/\rho_0}} \sim \Pi_1^{-(1-\alpha)} . \qquad (10.47)$$

Finally, it turns out that if one constructs the distributions of density, pressure, and speed in reduced coordinates, taking x_f as the length scale and p_f, ρ_f and u_f as scales for the flow properties, then those distributions just as rapidly become independent of time (see Figure 10.4). In other words, it turns out that the solution of the problem rapidly approaches the self-similar asymptotics

$$\xi_f = \xi_0(\gamma)\Pi_1^\alpha , \qquad \Phi_\rho = \Phi_{1\rho}\left(\frac{\Pi_2}{\Pi_1^\alpha}\right) ,$$

$$\Phi_p = \Pi_1^{-2(1-\alpha)}\Phi_{1p}\left(\frac{\Pi_2}{\Pi_1^\alpha}\right) , \qquad \Phi_u = \Pi_1^{-(1-\alpha)}\Phi_{1u}\left(\frac{\Pi_2}{\Pi_1^\alpha}\right) . \qquad (10.48)$$

We emphasize that the approach of the solution to this self-similar asymptotics does not occur uniformly in the whole region of motion $-\infty < x \le x_f(t)$ but only close to the front $x_f(t)$, in a region that increases with the time elapsed since the start of the expansion. Further, numerical calculation (Zhukov and Kazhdan, 1956) performed under the condition that $\rho_1 = 0$ (corresponding to expansion into a vacuum) showed that the solution approaches a self-similar asymptotics of the form (10.48) with one and the same exponent α independently of whether the pressure at the dividing wall during the time interval τ is constant or changes according to some law.

It is natural to try assuming the incomplete similarity of the intermediate asymptotics (see Chapter 9) and constructing the limiting self-similar solution directly. We seek it in following the class of solutions:

$$p = p_0 \frac{x_2}{t^2} P(\xi) , \qquad \rho = \rho_0 R(\xi) , \qquad u = \frac{x}{t} V(\xi) ,$$

$$x_f = At^\alpha , \qquad \xi = \frac{x}{At^\alpha} ; \qquad (10.49)$$

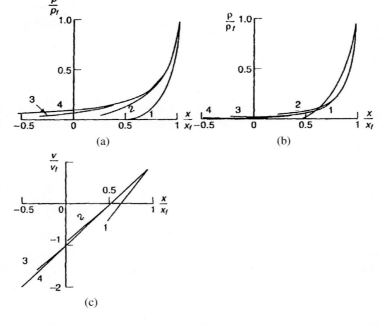

Figure 10.4 Dependences of (a) p/p_f, (b) ρ/ρ_f and (c) v/v_f on x/x_f, obtained by numerical solution of the non-self-similar impulsive load problem for $\gamma = 1.4$, which rapidly approach the dependences corresponding to a self-similar intermediate asymptotics (curve 4). From Zhukov and Kazhdan (1956) with kind permission from Springer Science+Business Media B.V.: 1, $\theta = 1.6$; 2, $\theta = 5/0$; 3, $\theta = 15.0$; 4, self-similar intermediate asymptotics.

here A and α are constants. For the functions P, V and R we obtain a system of ordinary differential equations, which reduce to the first-order equation

$$\frac{dz}{dV} = \frac{z}{\Delta} \left\{ [2(V-1) + (\gamma-1)V](V-\alpha)^2 \right.$$
$$\left. - (\gamma-1)V(V-1)(V-\alpha) - [2(V-1) + \kappa(\gamma-1)]z \right\}, \qquad (10.50)$$
$$\Delta = (V-\alpha)[V(V-1)(V-\alpha) + (\kappa-V)z],$$

with

$$\kappa = \frac{2(1-\alpha)}{\gamma}, \qquad z = \frac{\gamma P}{R}, \qquad (10.51)$$

and to two other first-order equations,

$$\frac{d \ln \xi}{dV} = \frac{z - (V-\alpha)^2}{V(V-1)(V-\alpha) + (\kappa-V)z} \qquad (10.52)$$

and

$$\frac{d \ln R}{d\xi} = -\frac{V(V-1)(V-\alpha) + (\kappa - V)z}{(V-\alpha)[z-(V-\alpha)^2]} - \frac{V}{V-\alpha}. \tag{10.53}$$

Thus, if the desired solution of (10.50) is known, the solutions of (10.52) and (10.53) can be found by quadratures. This desired solution must pass through two singular points, the image of the shock front

$$V = \frac{2\alpha}{\gamma+1}, \qquad z = \frac{2\alpha^2\gamma(\gamma-1)}{(\gamma+1)^2}, \tag{10.54}$$

and the image of the free boundary, the singular point

$$V = \kappa, \qquad z = \infty. \tag{10.55}$$

It is easy to show that this singular point is of saddle type. The variable ξ must increase monotonically as one moves from the singular point (10.55) to the image of the shock front, (10.54). Thus the mathematical problem turns out in this case to be a nonlinear eigenvalue problem, that of obtaining an integral curve, for a first-order equation, passing through two points, one of which is a saddle-type singular point. This is impossible in general but one can show that for each γ there exists a value of α, an *eigenvalue* of the problem, for which the integral curve of (10.50) passing through the image of the shock front also goes through the saddle, the image of the free boundary. An investigation of the field of integral curves of equation (10.50) was performed by V. A. Adamsky (see Adamsky, 1956). The values of α corresponding to various values of γ over its complete range $1 \leq \gamma \leq \infty$ are given in the table below. It can be seen that for all γ in the interval $1 < \gamma < \infty$ we have the inequality $\frac{1}{2} < \alpha < \frac{2}{3}$.

γ	1.0	1.1	7/5	5/3	2.8	∞
α	1/2	0.569	3/5	0.611	0.627	0.64

It turns out that values of the exponent α determined by direct construction of a limiting self-similar solution agree well with the values obtained by numerical calculation of an asymptotic solution of the non-self-similar problem performed by A. I. Zhukov and Ya. M. Kazhdan (see Zhukov and Kazhdan, 1956).

Evidently the limiting self-similar solution (10.49) is determined by direct construction only to within a constant A; comparison of (10.49) and (10.46) gives

$$A = \xi_0(\gamma) \sqrt{p_w/\rho_0} \, \tau^{1-\alpha}. \tag{10.56}$$

Thus, to obtain the same asymptotics while reducing the duration τ of the impulse acting on the gas, we must correspondingly increase the pressure at the dividing wall according to the law

$$p_w = \text{const} \times \tau^{-2(1-\alpha)} . \tag{10.57}$$

Now we will discuss a seeming paradox concerning the asymptotic solution obtained. The mass of gas involved at each moment in the flow through unit area of the boundary is finite. Hence the laws of conservation of momentum and energy apply; these are also valid at the non-self-similar stage of the motion. Therefore, the idea naturally occurs of using these laws to determine the exponent α and the constant A of the limiting self-similar solution.

The gas is initially at rest and at zero pressure, so its momentum and energy are zero. The total momentum J of the gas involved in the motion is equal at any instant to the impulse of the pressure load:

$$J = \beta p_w \tau , \qquad \beta = \int_0^1 f(\lambda)\, d\lambda . \tag{10.58}$$

Hence we obtain the momentum conservation law in the form

$$\beta p_w \tau = \int_{-\infty}^{x_f} \rho u \, dx . \tag{10.59}$$

As time increases the solution tends to a self-similar one. Hence, it might appear that if we pass to the limit under the integral sign then we can substitute into (10.59) the expressions for the density and speed from the self-similar solution (10.49) and obtain the relation

$$\beta p_0 \tau = \rho_0 A^2 t^{2\alpha-1} \int_{-\infty}^1 R(\xi) V(\xi) \, d\xi . \tag{10.60}$$

Since the integral on the right-hand side is obviously independent of time it is necessary, in order for the left-hand side also to be independent of time, that the relation $\alpha = 1/2$ is satisfied, after which it would appear that one could find the constant A from (10.60).

However, we also have the energy conservation law. According to this, the work per unit area performed by the loading due to the gas is equal to

$$\int_0^\tau p(0,\tau) u(0,t)\, dt = \delta p_w^{3/2} \rho_0^{-1/2} \tau , \tag{10.61}$$

where δ is a numerical constant. But the energy of the gas about to enter the motion is zero because its speed and pressure are equal to zero. Hence the energy of the gas actually involved in the motion is at any instant equal to the work performed by the impulsive loading:

$$\delta p_w^{3/2} \rho_0^{-1/2} \tau = \int_{-\infty}^{x_f} \rho \left[\frac{u^2}{2} + \frac{p}{(\gamma - 1)\rho} \right] dx . \tag{10.62}$$

Again, it seems that passing to the limit under the integral sign and substituting the expressions for speed, density and pressure from the limiting self-similar solution, we obtain

$$\delta p_w^{3/2} \rho_0^{-1/2} \tau = \rho_0 A^3 t^{3\alpha - 2} \int_{-\infty}^{1} R \left[\frac{V^2}{2} + \frac{P}{(\gamma - 1)R} \right] \xi^2 d\xi . \tag{10.63}$$

At first glance it would seem to follow from this that $\alpha = 2/3$ and that (10.63) also allows one to determine the constant A. Thus a paradox arises, consisting of the fact that the exponents α in the self-similar variable determined from the laws of conservation of momentum ($\alpha = 1/2$) and of energy ($\alpha = 2/3$) do not agree with each other or with the range of values of the exponent α ($1/2 < \alpha < 2/3$) determined by direct construction of a limiting self-similar solution or by its numerical calculation.

The resolution of this paradox is simple and at the same time instructive. The fact is that the integral in the momentum equation (10.60) is equal to zero, and the integral in the energy equation (10.63) is equal to infinity, so that from these relations it is impossible to determine either the exponent α or the constant A. The transition to the limit under the integral sign in the conservation laws (10.59) and (10.62) was itself inadmissible, because the convergence of the integrands to the limit is non-uniform over the domain of integration. Usually we leave such fine points as uniform convergence under the integral sign to pure mathematicians, assuming that everything will be in order. Here it is not so, and that is the essence of the problem!

In fact the limiting self-similar motion is obtained by transition to the limit over the entire domain $-\infty < x \leq x_f$, with the duration τ of the impulse tending to zero and the pressure on the boundary tending to infinity according to the law $p_w = \text{const} \times \tau^{-2(1-\alpha)}$. Here the total momentum $\beta p_w \tau$ tends to zero as the factor const and the energy $\delta p_w^{3/2} \rho_0^{-1/2} \tau$ to infinity as const $\times \tau^{2\alpha - 1} \times \tau^{3\alpha - 2}$, so that (we recall that α lies between $1/2$ and $2/3$) the self-similar limiting motion has zero momentum and infinite energy. Further, the self-similar solution is limiting to the solution of the original non-self-similar problem with finite p_w and τ and with t tending to infinity. However, as has already been mentioned, the convergence to the limiting solution is non-uniform in the domain $-\infty < x \leq x_f$. The momentum of the region of compression $x_0(t) \leq x \leq x_f(t)$ grows infinitely with time. The momentum of the region of expansion $-\infty < x \leq x_0(t)$ has a negative sign and its absolute value also grows infinitely with time. The algebraic sum of these two

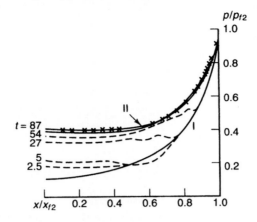

Figure 10.5 The solution to the problem of an intense explosion at an interface settles down to an intermediate asymptotics, at first internal (I) and later external (II). The crosses correspond to $\theta = t/\tau = 100$ (see (10.45)). From Vlasov, Derzhavina and Ryzhov (1974) with kind permission from Springer Science+ Business Media B.V.

momenta, equal to $\beta p_w \tau$, becomes ever smaller compared with the momentum of each of the two regions mentioned; it is different from zero only because of the departure of the motion from self-similarity.

We now consider the energy $\mathcal{E}$ of any region $x_1(t) = \xi_1 A t^\alpha \le x \le x_f(t)$ in which the motion becomes close to self-similar starting from some instant of time:

$$\mathcal{E} = \int_{x_1}^{x_f} \rho \left[\frac{u^2}{2} + \frac{p}{(\gamma - 1)\rho} \right] dx = \rho_0 A^3 t^{3\alpha - 2} \int_{\xi_1}^{1} R \left[\frac{V^2}{2} + \frac{P}{(\gamma - 1)R} \right] \xi^2 \, d\xi \, . \tag{10.64}$$

It is evident that the energy $\mathcal{E}$ tends to zero with increasing t, so that the contribution of the self-similar region to the bulk energy becomes ever less in time and the basic contribution to the energy is determined by the motion close to the left-hand boundary, where it always remains non-self-similar no matter how much time has passed since the start of the motion.

The ultimate step in the investigation of the problem of an intense explosion at an interface, summing up the results of von Weizsäcker (1954), Zeldovich (1956) and their associates and Nigmatulin (1965), was a numerical computation performed by O. S. Ryzhov and his associates (Vlasov, Derzhavina and Ryzhov, 1974). Figure 10.5 demonstrates the transition from the internal asymptotics (I) to the external asymptotics (II) for the following initial conditions:

$$
\begin{aligned}
u = 0 \, , \qquad & \rho = \rho_1 = 0.1 \, , \qquad && p = 0 \qquad && (x < 0) \, ; \\
u = 0 \, , \qquad & \rho = \rho_0 = 1 \, , \qquad && p = 1 \qquad && (0 < x < 1) \, ; \\
u = 0 \, , \qquad & \rho = \rho_0 = 1 \, , \qquad && p = 0 \qquad && (x > 0) \, .
\end{aligned}
\tag{10.65}
$$

It is seen that the internal intermediate asymptotics is established close to the shock wave with sufficient accuracy for $\theta_1 < \theta < \theta_2$, where θ_1 is only 2.5. After $\theta = \theta_2 \sim$ 40 the finiteness of the density on the left-hand side of (10.64) becomes substantial and at $\theta \sim 100$ the external intermediate asymptotics is established; this will hold until the pressure at the right-hand shock front becomes comparable with the initial pressure.

11

Turbulence: generalities; scaling laws for shear flows

Turbulence is the state of vortex fluid motion where the velocity, pressure and other properties of the flow field vary in time and space sharply and irregularly and, it can be assumed, randomly.

Turbulent flows surround us, in the atmosphere, in the oceans, in engineering systems and sometimes in biological objects. A contrasting class of fluid motions, when the fluid moves in distinguishable layers (*laminae* in Latin) and the flow-field properties vary smoothly in time and space, is known as *laminar* flow. In Figure 11.1 an example of the time dependence of the velocity in a turbulent flow is presented. For laminar flow the time dependence would be a smooth line.

Leonardo da Vinci[1] already knew about and clearly distinguished these two types of flow. Leonardo even used the term "turbulenza". However, his observations and thoughts were buried in his notebooks, solemnly and carefully preserved in the Royal and Papal archives. They were not published until recently and therefore most regrettably did not influence future studies.

The systematic scientific study of turbulence began only in the nineteenth century, and here two names should be mentioned in particular, those of the French applied mathematician Joseph Boussinesq and the British physicist Osborne Reynolds. Boussinesq was a student of A. Barré de Saint Venant, who, in his turn, was a devoted disciple of C. L. M. H. Navier, the originator of the mathematical models for both Newtonian viscous fluid flows and the deformation of perfectly elastic bodies.

It was Boussinesq (1877) who discovered the basic mechanism of the transverse transfer of momentum in turbulent flows. In contrast with laminar flow this is due not to molecular diffusion but to the action of vortices: turbulent flow is "stuffed" with vortices and they are wholly responsible for enhancing the transfer of momentum in turbulent flows, as well as for the randomness of turbulent flow. Boussinesq

---

[1] The author is obliged to Academician U. Frisch for this historical reference.

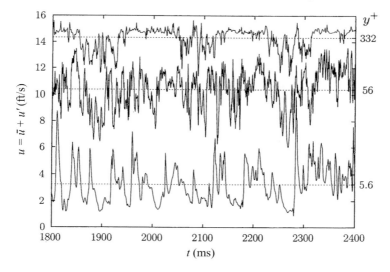

Figure 11.1 An example of the time dependence of the velocity in a turbulent flow: the longitudinal component of the velocity u in a turbulent boundary layer at three dimensionless distances from the wall, 5.6, 56, and 332. From Wu and Moin (2009).

introduced the concept of the "eddy viscosity" (*viscosité tourbillonaire*), quantifying the intensity of the momentum transfer in turbulent flows by vortices.

It was Reynolds (1883, 1894), who, by visualization, investigated the transition to turbulence in pipes with transparent walls and formulated the fundamental criterion of transition. A careful analysis of the memoir of Reynolds (1883) (see also the illuminating article of Sir James Lighthill, 1970) shows that he clearly understood that his criterion for the transition to turbulence in pipes,

$$Re = \frac{\bar{u}d}{\nu} = \text{const}$$

(where $\bar{u}$ is the average velocity, the fluid flux through a cross-sectional area divided by that area, d is the pipe diameter, ν is the fluid's kinematic viscosity), is not a universal criterion. Indeed, he formulated this criterion in the following way: "It seems, however, to be certain *if the eddies are due to one particular case* [in modern language, if all other factors are fixed; the italicization is that of the present author], that integration would show the birth of eddies to depend on some definite value of $d\rho\bar{u}/\eta$." Here, the notation has been changed to ours and "*one particular case*" can be interpreted as meaning that it is so for given initial conditions at the entrance to the pipe, the design of the entrance etc. It can be added that recently a very instructive series of experiments was performed by Professor A. A. Paveliev of the Moscow State University. His set-up contained a diaphragm at the entrance to the pipe in which there were 250 orifices; the fluid was supplied to each of them independently, keeping the total flux fixed. The transition Reynolds number was found to be different for different distributions of fluxes.

Moreover, about a hundred years after Reynolds' work, N. J. Johannesen and C. Lowe repeated his experiments (see van Dyke, 1982, p. 61) on the same set-up, which had been carefully preserved by Manchester University. They obtained a transition Reynolds number much less than the value, 13 000, observed by Reynolds. Johannesen and Lowe attributed this reduction to the traffic on the streets. However, Ekman (1910) and Schiller (1922) reported experiments where the water had been calm for years and the entrance to the pipe was especially smooth. In these experiments, laminar flows were obtained for *Re* up to 40 000. Now let us imagine an ideal experiment, in which the influence of the ambient media is prevented and an ideal laminar velocity profile is produced at the entrance to the pipe. In spite of all these precautions, certain sources of perturbation still remain that can influence the transition and that are non-universal. These are the thermal oscillations of the molecules and atoms of which the fluid is composed and the fact that the development of vortices along the pipe is non-monotonic, so that at first their intensity can decrease but later, if the pipe is sufficiently long, they will start to grow.

So, in this ideal experiment, the transition Reynolds number depends on two dimensionless parameters,

$$\Pi_1 = \frac{\sqrt{\overline{v'^2}}}{\overline{u}}, \qquad \Pi_2 = \frac{\ell}{d};$$

here $\overline{v'^2}$ is the mean square of the thermal oscillation velocities, averaged over frequency and over the volume of the elementary fluid particle, ℓ is the length of the pipe and d is the diameter. The quantity $\overline{v'^2}$ can depend on the following governing parameters: the fluid density ρ_0, the volume w of the elementary fluid particle, the Boltzmann constant k_B and the temperature θ. Their dimensions are as follows: $[\rho] = M/L^3$, $[\theta] = \Theta$, $[k_B] = ML^2T^{-2}\Theta^{-1}$, $[w] = L^3$. All these dimensions are independent, so standard dimensional analysis gives $\overline{v'^2} = \text{const} \times k_B\theta/(\rho_0 w)$ (the derivation of this formula is due to Boris Ya. Zeldovich). Thus $\Pi_1 = \text{const} \times k_B\theta/(\rho_0 w \overline{u})$.

This formula suggests an interesting way of obtaining the laminar flow of super-cooled helium at very large Reynolds numbers: as P. L. Kapitsa discovered, the viscosity of supercooled liquid helium decreases dramatically.

However, the general belief that turbulent flows cannot exist for *Re* < 2000 remains unproven experimentally, because, in a laboratory, the possible pipe length might be insufficient.

There is a general belief, also unsupported by rigorous mathematical proof, that laminar flow in pipes is stable to infinitesimally small perturbations. So, on the whole, the transition to turbulence in pipes still remains an open problem.

Reynolds (1894) proposed a general approach for studying random turbulent flows by means of averaging, and formulated the basic rules for this. Reynolds called turbulent flows "sinuous"; the term "turbulent flows" was reintroduced, independently of Leonardo da Vinci, by Lord Kelvin – a remarkable example of congeniality!

For the past hundred years turbulence has been studied by an army of engineers, mathematicians and physicists, including such giants as A. N. Kolmogorov, W. Heisenberg, G. I. Taylor, L. Prandtl and Th. von Kármán. Every advance in a wide collection of subjects, from chaos and fractals to field theory, and every increase in the speed and parallelization of computers has been heralded as ushering in the solution of the "turbulence problem". Nevertheless turbulence, the fatigue of structures (in a wider sense, including plasticity), the friction of solid bodies and the fracture of structures – problems of comparable practical value and fundamental interest – remain a great challenge. In particular, almost nothing is known about turbulence, even in incompressible Newtonian fluids, from first principles, i.e. from the Navier–Stokes and continuity equations. This provides an important and interesting intellectual puzzle for future researchers. Meanwhile it is doubtless true that turbulent flows of wide classes of fluids, including air and water under normal conditions, can be described by the Newtonian viscous fluid approximation down to the scale of the smallest vortices and beyond.

At the end of his life A. N. Kolmogorov, whose ideas shaped modern turbulence studies, said, surveying the beginning of his work on turbulence:

> It became clear for me that it is unrealistic to have a hope for the creation of a pure theory [of turbulent flows of fluids and gases, based on first principles Q] closed in itself. Due to the absence of such a theory we have to rely upon the hypotheses obtained by the processing of experimental data ...

Unfortunately very little, if anything, has changed since these words were said regarding the possibility of constructing a pure closed theory. However, more experimental data have appeared and it is possible now, where needed, to modify previous hypotheses and the special models based on these hypotheses. Examples of such modifications will be presented below.

11.1 Kolmogorov's example

At the beginning of his course on turbulence, delivered at Moscow State University in 1954, Kolmogorov asked listeners the following question: what would be the velocity on the surface of the River Volga (in Russia; its parameters are close to those of the Mississippi River in the USA) if by some miracle the flow in the river, *preserving its geometry*, became laminar?

To answer this question, Kolmogorov naturally modeled the river by a weakly inclined (with slope $i \ll 1$) spatially homogeneous open channel (Figure 11.2a). In this simple case of laminar flow in a channel, the system of Navier–Stokes equations is reduced to a single equation,

$$\eta \frac{d^2u}{dz^2} + \rho g i = 0 . \tag{11.1}$$

Here $u(z)$ is the velocity parallel to the bottom, z is the coordinate reckoned from the bottom and perpendicular to it, η is the water's dynamic viscosity, ρ is its density and g is the gravity acceleration. The natural boundary conditions are $u(0) = 0$ (a no-slip condition at the bottom) and $du/dz = 0$ at $z = H$, with H the river depth (a no-shear condition at the river surface). The easily obtainable solution is

$$u(z) = \frac{\rho g i H^2}{2\eta} \left(\frac{2z}{H} - \frac{z^2}{H^2} \right) . \tag{11.2}$$

Thus, the velocity u_{surf} at the river surface would appear to be given by

$$u_{\text{surf}} = \frac{\rho g i H^2}{2\eta} . \tag{11.3}$$

Now, let us estimate u_{surf} using realistic parameter values: $\eta/\rho = 10^{-2}$ cm^2/s, $H = 20$ m $= 2 \times 10^3$ cm, $i = 10^{-4}, g = 10^3$ cm/s^2. The obviously absurd value $u_{\text{surf}} = 2 \times 10^7$ cm/s $= 200$ km/s $\cong 400\,000$ miles/hour is obtained. The reason for this absurdity is that the flow in the river is not laminar, but turbulent; it is, as we described earlier, "stuffed" with vortices (Figure 11.2b). The "eddy viscosity" η_{turb}, in contrast with the molecular viscosity, is *no longer a fluid property*. It is a *local flow property*, different in different places. The value η_{turb} needed to obtain a realistic value from (11.3) at the river surface exceeds the molecular viscosity η two hundred thousand times!

Kolmogorov's example is especially significant, even fundamental, in the following respect: it clearly demonstrates the huge reserves of energy available in natural flows. Natural flows would release part of their hidden energy if, in some way, even partial flow laminarization were achieved. In such situations the eddy viscosity would reduce and the flow velocity would increase, sometimes substantially. This is what happens in reality: partial laminarization is achieved in dust storms (the laminarizing factor is the presence of suspended dust particles), tropical hurricanes (the laminarizing factor is the presence of water droplets formed when the water waves at the ocean surface break), firestorms (the laminarizing factor is the burning debris and soot particles) and some other natural phenomena.

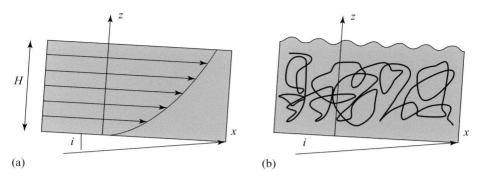

Figure 11.2 Kolmogorov's example. (a) Laminar flow in a channel. (b) Turbulent flow in a river.

11.2 The Reynolds equation. Reynolds stress

The "realizations" of turbulent flows under prescribed external conditions, e.g. a given pressure drop at the ends of a pipe, are highly unstable; their measurement is impossible, in the first place because even delicate measuring devices will change these individual motions. Therefore turbulence studies operate with averages of the fluid flow properties. In theoretical studies "ensemble" or "probability" averaging is used: averaging is made over the whole ensemble of possible turbulent flow realizations under prescribed external conditions.

This way of averaging has the following properties (a bar denotes an average value):

$$\overline{f + g} = \bar{f} + \bar{g}, \tag{11.4}$$

$$\overline{af} = a\bar{f}, \qquad \text{if } a = \text{const}, \tag{11.5}$$

$$\bar{a} = a, \qquad \text{if } a = \text{const}, \tag{11.6}$$

$$\overline{\frac{\partial f}{\partial s}} = \frac{\partial \bar{f}}{\partial s}, \tag{11.7}$$

where s is a spatial or time coordinate, and

$$\overline{\bar{f}g} = \bar{f}\bar{g}. \tag{11.8}$$

The conditions (11.4)–(11.8) are generally known as *Reynolds' rules* of averaging. Thus, corresponding to every component of the turbulent flow field f an average value $\bar{f}$ is assigned, and its random fluctuation is $f' = f - \bar{f}$.

We will apply these rules to the averaging of the Navier–Stokes equations. We take the momentum balance equation in Cartesian orthonormal coordinates and average it:

$$\overline{\partial_t \rho u_i} + \overline{\partial_\alpha \rho u_i u_\alpha} = -\overline{\partial_i p} + \overline{\partial_\alpha \tau_{i\alpha}} \,. \tag{11.9}$$

Here $\tau_{i\alpha}$ is a component of the viscosity stress tensor. Using the rules of averaging (11.4)–(11.8) we obtain

$$\overline{\partial_t \rho u_i} = \rho \partial_t \bar{u}_i \,;$$

$$\overline{\partial_\alpha \rho u_i u_\alpha} = \overline{\partial_\alpha \rho (\bar{u}_i + u_i')(\bar{u}_\alpha + u_\alpha')}$$

$$= \partial_\alpha \rho \bar{u}_i \bar{u}_\alpha + \partial_\alpha \overline{\rho \bar{u}_i u_\alpha'} + \partial_\alpha \overline{\rho u_i' \bar{u}_\alpha} + \partial_\alpha \overline{\rho u_i' u_\alpha'}$$

$$= \partial_\alpha \rho \bar{u}_i \bar{u}_\alpha + \partial_\alpha \overline{\rho u_i' u_\alpha'} \,,$$

because, according to the rule (11.8),

$$\partial_\alpha \overline{\rho u_i' \bar{u}_\alpha} = \partial_\alpha \overline{\rho \bar{u}_i u_\alpha'} = 0 \,.$$

Thus, the averaged Navier–Stokes equation (11.9) can be represented in the form

$$\rho \partial_t \bar{u}_i + \rho \bar{u}_\alpha \partial_\alpha \bar{u}_i = -\partial_i \bar{p} + \partial_\alpha (\bar{\tau}_{i\alpha} + T_{i\alpha}) \,. \tag{11.10}$$

Here

$$T_{i\alpha} = -\rho \overline{u_i' u_\alpha'} \tag{11.11}$$

is obviously a component of the *tensor*

$$\mathbf{T} = -\rho \overline{\mathbf{u}' \otimes \mathbf{u}'} \tag{11.12}$$

which represents the rate of momentum transfer by turbulent vortices. In laminar flows $\mathbf{u}' \equiv 0$, and the tensor $\mathbf{T}$ vanishes. It is instructive that the tensor $\mathbf{T}$ enters the averaged Navier–Stokes equation as an addition to the averaged viscous stress $\bar{\tau}$:

$$\partial_t \rho \bar{\mathbf{u}} + \rho (\bar{\mathbf{u}} \cdot \nabla) \cdot \bar{\mathbf{u}} = -\nabla \bar{p} + \nabla \cdot (\bar{\tau} + \mathbf{T}) \,. \tag{11.13}$$

Equation (11.10) (in vector form (11.13)) was obtained by Reynolds and is traditionally named the *Reynolds equation* after him. Accordingly the tensor $\mathbf{T}$ is named the *Reynolds stress tensor.*

The formidable difficulty of the turbulence problem lies exactly in the fact that nothing is known in general about the Reynolds stress tensor.

This renders unclosed the system comprising the Reynolds equation (11.10) – or, in vector form, (11.13) – and the obviously obtainable averaged continuity equation

$$\nabla \cdot \bar{\mathbf{u}} = 0 \,, \qquad \text{or} \qquad \partial_\alpha \bar{u}_\alpha = 0 \,. \tag{11.14}$$

The problem of turbulence is in fact the problem of finding the Reynolds stress tensor $\mathbf{T}$, i.e. finding its connection with the average components of the turbulent flow field.

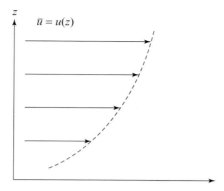

Figure 11.3 Shear flow.

11.3 Turbulent shear flow

Shear flow is statistically steady, homogeneous in the direction of the average velocity flow and such that all average flow properties vary only in the direction perpendicular to the direction of the mean velocity (Figure 11.3).

Studies of turbulent shear flows are of special significance from both theoretical and experimental viewpoints. Indeed, for turbulent shear flows it is possible to advance deeper without accepting additional hypotheses that are sometimes doubtful; their mathematical models are substantially simplified. The main reason for such simplifications is that general turbulent flow fields are non-local, both in time and space. This means that the average flow properties at a certain point and at a certain moment in time depend not only on the properties of the flow at this point but also on the properties in a neighborhood around the point under consideration and in the time interval before the moment considered.

This is not the case for turbulent shear flows. The average flow field properties at a point can be considered, for a shear flow, as local properties depending only upon the flow properties at the same point. The reason for such a simplification is that in the averaging process the contributions of the neighborhood and of the prehistory vanish owing to the steadiness and homogeneity of the mean flow in the direction of the mean velocity: positive and negative fluctuations compensate each other. Also, an important advantage of shear flows from the viewpoint of experimentalists is the "ergodicity" property, the possibility of calculating ensemble averages by averaging at a certain point over a time interval (this is allowable owing to the steadiness of the flow) or over a longitudinal space interval (this is allowable owing to the spatial homogeneity of the averaged field in the mean flow direction).

11.4 Scaling laws for turbulent flows at very large
Reynolds numbers. Flow in pipes

Turbulence at very large Reynolds numbers ("very large" means that we have not only $Re \gg 1$ but also $\log Re \gg 1$) is often called "developed turbulence" and is widely considered to be one of the happier provinces of the turbulence realm, as it is thought that two of its basic results are well established and should enter, basically untouched, into a future complete theory of turbulence. These results are the *von Kármán–Prandtl universal logarithmic law* in the wall region of wall-bounded turbulent flow and the *Kolmogorov–Obukhov scaling laws* for the local structure of developed turbulent flows.

The start of fundamental research into turbulent shear flows at very large Reynolds numbers can be dated precisely from the lecture of Th. von Kármán at the Third International Congress for Applied Mechanics at Stockholm, 25 August 1930.

The second major breakthrough in the theory of turbulence at very large Reynolds numbers occurred in 1941, in the fundamental works of A. N. Kolmogorov and of A. M. Obukhov, at that time Kolmogorov's graduate student (Kolmogorov, 1941a, b; Obukhov, 1941a, b), where the scaling laws for the local structure of such flows were obtained. The role of an elucidating article by Batchelor (1947) should be emphasized; in this article the Kolmogorov–Obukhov theory, presented originally in the form of short notes, was explained in detail and fundamentally clarified. It happened that even Russian students (including the present author) learned the theory from reading this article. The Kolmogorov–Obukhov theory will be discussed in the next chapter.

Von Kármán was one of the principal founders of the International Congresses for Applied Mechanics. Unquestionably his lecture "Mechanical similitude and turbulence" was the central event of the Congress. Von Kármán began his lecture (von Kàrmàn, 1930) with the following statement:

> Our experimental knowledge of the internal structure of turbulent flows is insufficient for delivering a reliable foundation for a rational theoretical calculation of the velocity distribution and drag in the so-called hydraulic flow state. Numerous semi-empirical formulae, for instance, the attempt to introduce turbulent drag coefficients, are unable to satisfy either the theoretician or the practitioner. The investigations which will be presented below also do not claim to achieve a genuine ultimate theory of turbulence. I will restrict myself rather to clarifying what can be achieved on the basis of pure fluid dynamics if definite hypotheses are introduced concerning definite basic questions.

The hypothesis proposed by von Kármán for answering the fundamental questions concerning the velocity distribution and drag coefficients in turbulent hydraulic flows or, as they are called now, shear flows – primarily, flows in pipes and channels – was presented by him in the following straightforward form.

> On the basis of these experimentally well-established facts we make the assumption that away from the close vicinity of the wall the velocity distribution of the mean flow is viscosity independent.

As a result of subsequent arguments proposed by von Kármán there appeared what is called now the universal (i.e. Reynolds-number-independent) *logarithmic law* and the corresponding drag law for turbulent flow in a cylindrical pipe. These will be presented below.

The leaders in applied mechanics of that time were present at von Kármán's lecture and took part in the subsequent, very impressive, discussion. The first speaker was Ludwig Prandtl. He said:

> The new Kármán calculations signify very pleasing progress in the problem of fluid friction. It was always the case that by advancing to higher Reynolds numbers the previous interpolation formulae were revealed to be incorrect by extrapolation to a newly investigated range and had to be replaced by new ones. Research laboratories made big efforts to achieve higher Reynolds numbers, but the cost of big experimental set-ups has a bound which cannot be substantially exceeded. *Due to Kármán's formulae further efforts in this direction became unnecessary* [present author's italics]. The formulae are in such good agreement with the experiments in pipe flows by Nikuradze, and by Schiller and Hermann, and with experiments concerning the drag of plates performed by Kempf, that complete confidence can be placed in them for their application at arbitrarily large Reynolds numbers. *For lower Reynolds numbers the agreement is worse, and this can be attributed to the action of the viscosity also in the inner part of the flow, i.e. to the viscosity-influenced streaks of which the laminar layer at the wall consists and which in this case enter far into the internal part of the flow* [present author's italics].
>
> I want to point out a seeming contradiction concerning the representation of the velocity distribution by Nikuradze in connection with Kármán's new formulae and my earlier formulation using the dimensionless distance from the wall. Kármán's formulae use viscosity in the boundary condition only. The velocity distribution should be calculated without viscosity. However, the dimensionless distance from the wall, $y^* = (y/\nu)\sqrt{\tau_0/\rho}$, does contain the viscosity. *According to my opinion, the explanation is that the Kármán representation should be considered as exact for very large Reynolds numbers, while the representation via the dimensionless distance from the wall applies essentially to the wall layer and streaks where the viscosity and turbulence are acting together* [present author's italics].

It should be understood that at that time Prandtl was generally considered as "the chief of applied mechanics" (see Batchelor, 1996, p. 185). The opinion which we have just reproduced explains at least partially why over nearly 70 years the Nikuradze (1932) experiments were never extended to larger Reynolds numbers.

Moreover, the culture of such experiments, in fact very subtle, decayed and to a certain extent was lost.

It is also true that the last part of Prandtl's comment is very deep and instructive, but it remained dormant and was not cast into a proper mathematical theory for the following technical reason. In the early 1930s, and even long before, the mathematical techniques which were needed here were in sufficiently good shape. However, they were considered rather as mathematical monsters with no practical application. Only several decades later was it recognized (see Chapter 9) that these techniques were needed for the modeling of many physical phenomena; such techniques then entered the practice of applied mathematics and theoretical physics as incomplete similarity, fractals and renormalization groups. These concepts will be used in the present chapter to explain the situation regarding the scaling laws for turbulent shear flows at very large Reynolds numbers. In particular, the concept of incomplete similarity will allow a resolution of the contradiction mentioned in the second paragraph of Prandtl's comment.

We consider here the scaling laws for wall-bounded shear flows at very large Reynolds numbers. Among such flows are many flows of practical importance: flows in pipes, channels, wall-jets, and boundary layers. Our presentation is based on a series of works by A. J. Chorin, V. M. Prostokishin and the present author (see the reviews of Yaglom, 1993, 2000; Barenblatt, Chorin and Prostokishin, 1997, 2000; Chorin, 1998; and Barenblatt, 2003; and the sources given therein).

In particular, flows in cylindrical pipes (Figure 11.4) constitute an instructive example of wall-bounded turbulent shear flows. The advantage of considering these flows here is the availability of a large amount of experimental data: it allows the clarification of even fine details of the turbulent shear flow structure.

Thus, we have the same clear goal and well-determined problems as those once formulated by von Kármán (see above): to obtain mathematical expressions for the drag coefficient and the velocity distribution in the intermediate region 3 (see Figure 11.4) of the flow in a pipe at very large Reynolds numbers. "Intermediate" means here the region between the viscous sublayer 1 adjacent to the wall, where the velocity gradient is very high and the mean viscous stress is comparable with the Reynolds stress, and the region 2 close to the pipe axis. Von Kármán considered the same intermediate region, where the pipe flow carries its basic fluid discharge.

However, our basic hypothesis will be essentially different. In fact, *we reject von Kármán's hypothesis of complete viscosity independence and propose instead an hypothesis of incomplete similarity. This difference will lead to substantially different results whose agreement with experiment is instructive.*

Consider for comparison the traditional derivation of the velocity distribution in the intermediate region. The mean velocity gradient du/dz in a shear flow bounded

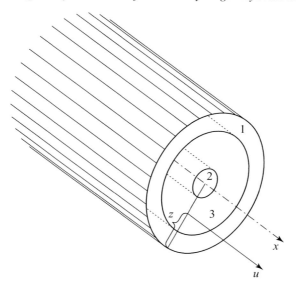

Figure 11.4 Flow in a long cylindrical pipe. The structure at large Reynolds number: 1, viscous sublayer; 2, near-axis region; 3, intermediate region. From Barenblatt (2003).

by a smooth wall, in particular, a flow in a pipe with smooth walls, could be governed by the following parameters: the transverse coordinate z (the distance from the wall); the shear stress at the wall τ; the external length scale entering the definition of the Reynolds number, in this case the pipe diameter d; and the fluid properties, its kinematic viscosity ν and density ρ. The velocity gradient du/dz is considered rather than the velocity itself, because the values of the velocity are influenced by the flow in the vicinity of the wall, region 1, where the intermediate asymptotic assumptions used in this derivation, as well in the modified derivation to be presented later, are invalid. Thus, for such a pipe flow,

$$\frac{du}{dz} = f(z, \tau, d, \nu, \rho) \,. \tag{11.15}$$

Introduce the viscosity length scale δ:

$$\delta = \frac{\nu}{u_*} \,, \qquad u_* = \sqrt{\frac{\tau}{\rho}} \,. \tag{11.16}$$

The quantity u_* is named the dynamic, or friction, velocity; its physical sense and importance will be explained later. A standard application of dimensional analysis gives

$$\frac{du}{dz} = \frac{u_*}{z} g\left(\frac{z}{\delta}, \frac{d}{\delta}\right) \,, \tag{11.17}$$

where g is a dimensionless function of its dimensionless arguments. Dimensional analysis also suggests that $d/\delta = u_*d/\nu$ is a function of the traditional Reynolds number $Re = \bar{u}d/\nu$, where $\bar{u}$ is the average velocity, i.e. the total discharge (fluid flux) divided by the cross-sectional area of the pipe. Therefore the relation (11.17) can be rewritten in the form

$$\frac{du}{dz} = \frac{u_*}{z} \, \Phi\left(\frac{z}{\delta}, Re\right), \qquad Re = \frac{\bar{u}d}{\nu}, \qquad (11.18)$$

where Φ is another dimensionless function of dimensionless arguments.

For very large Reynolds numbers in the intermediate region 3 considered, the ratio of the distance from the wall and the viscosity length scale z/δ is large. According to the basic von Kármán assumption (see the von Kármán quote above), the viscosity does not affect the velocity distribution in the intermediate region. However, in the expression (11.18) the viscosity ν enters both arguments since $z/\delta = u_*z/\nu$ and $Re = \bar{u}d/\nu$. Therefore the von Kármán assumption implies the complete independence of the parameters z/δ and Re. If so, according to a traditional (for teachers) argument the function Φ in the intermediate region can be replaced by a constant: $\Phi = 1/\kappa$. The constant κ was later named the "von Kármán constant". Substituting $\Phi = 1/\kappa$ into (11.18) gives

$$\frac{du}{dz} = \frac{u_*}{\kappa z}. \qquad (11.19)$$

Integration of (11.19) leads to a universal, Reynolds-number-independent, law for the velocity distribution:

$$u = u_*\left(\frac{1}{\kappa} \, \ln \frac{u_*z}{\nu} + C\right). \qquad (11.20)$$

We emphasize especially that in this line of argument the constant C is considered to be finite and Re-independent. This would seem to be logically consistent with the previous steps but is in fact a substantial additional hypothesis.

After the work of von Kármán (1930), Prandtl (1932) arrived at the universal logarithmic law (11.20) using a different approach, and so the term "von Kármán–Prandtl universal logarithmic law" became established. Many different derivations of this law have been proposed (e.g. Lighthill, 1970, pp. 116–17; Landau and Lifshitz, 1987, pp. 172–5; Schlichting, 1968, pp. 489–90; Monin and Yaglom, 1971, pp. 273–4; and, recently, Spurk and Aksel, 2008, pp. 219–23). We emphasize that the basis of all these derivations is the hypothesis explicitly formulated by von Kármán, quoted above.

Before the universal logarithmic law appeared, for practical needs engineers used power laws with exponents dependent on the Reynolds number. Von Kármán and Prandtl were successful in persuading the engineering world that this was incorrect and that the universal logarithmic law should be used instead.

However, for seven decades experimental information accumulated that suggested doubts regarding the universal logarithmic law. The analysis of new experimental data demonstrated a systematic deviation (not a scatter!) from the predictions of the universal logarithmic law even if a very liberal approach to the constants κ and C were allowed, although by the very logic of the derivation presented above the values of these constants should be identical for all high-quality experiments in smooth pipes. Moreover, an analysis by Schlichting (1968) of the experimental data of Nikuradze (1932), mentioned by Prandtl in his comment on the von Kármán lecture as a good confirmation of the von Kármán formulae, making future experiments unnecessary, suggested (see Figure 11.5) that power laws with Re-dependent powers were indeed a good representation of the experimental data concerning the velocity distribution in the intermediate region of pipe flow.

11.5 Turbulent flow in pipes at very large Reynolds numbers: advanced similarity analysis

Our knowledge of the Navier–Stokes equation and its solutions is at present insufficient to decide what kind of behavior takes place at large values of the function Φ governing the velocity gradient (equation (11.18)). However, analysis of the experimental data from investigations performed in the last two or three decades has shown that we do not have a case of complete similarity in the arguments of the function Φ and that the constant C entering the relation (11.20) is in fact strongly dependent on the Reynolds number.

Therefore it was a natural step to follow the general procedure described in Chapter 9 and to abandon the von Kármán assumption of complete viscosity independence.

The following basic hypothesis is assumed instead.

First hypothesis At large z/δ and very large Reynolds numbers there is incomplete similarity in the parameter $\eta = 1/\delta = u_* z/\nu$ and no similarity in the Reynolds number.

This means the following. At large $\eta = u_* z/\nu$ and very large Reynolds numbers Re the function $\Phi(\eta, Re)$ can be represented as a power monomial with coefficients dependent on the Reynolds number:

$$\Phi(\eta, Re) = A(Re)\eta^{\alpha(Re)} . \tag{11.21}$$

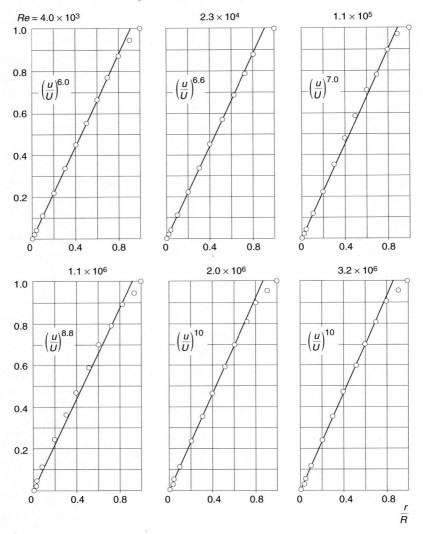

Figure 11.5 The experimental data of Nikuradze (1932) processed in the form of power laws $(u/U)^{n(Re)}$ versus r/R show that the power laws give a good approximation for the velocity distribution data. Here U is the maximum velocity, y the distance from the wall and R the pipe radius (Schlichting, 1968). The variation of n is significant: it goes from $1/6.0$ for $Re = 4.0 \times 10^3$ to $1/10$ for $Re = 3.2 \times 10^6$.

Here the coefficient A and the power $\alpha \neq 0$ in (11.21) are as yet undetermined functions of the Reynolds number Re.

Substituting (11.21) into (11.18), we obtain

$$\frac{du}{dz} = \frac{u_*}{z} A(Re)\eta^{\alpha(Re)} . \tag{11.22}$$

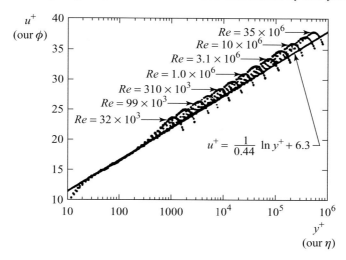

Figure 11.6 The Princeton data (Zagarola, 1996) obtained in a high-pressure pipe confirm the splitting of the experimental data according to their Reynolds numbers in the $(\ln \eta, \phi)$-plane $(\eta = u_* y / \nu, \; \phi = u/u_*)$. The solid line is the envelope; each dotted curve has its maximum at the η-value corresponding to the centre of the pipe. The splitting and form of the curves agree with the scaling law and are incompatible with the von Kármán–Prandtl universal logarithmic law.

Note immediately a clear-cut qualitative difference in the present case between complete and incomplete similarity. In the first case the experimental data should cluster in the traditional $(\ln \eta, \phi)$-plane (where $\phi = u/u_*$) on the single straight line of the universal logarithmic law. However, if one assumes incomplete similarity then the experimental points should occupy an area in the $(\ln \eta, \phi)$-plane: each value of the Reynolds number corresponds to a separate curve.

As Figures 11.5 and 11.6 demonstrate, experiments – even at high Reynolds numbers, performed at different times and in different laboratories – show a perceptible dependence of the distribution of the dimensionless velocity u/u_* on the Reynolds number. This splitting is incompatible with the universal von Kármán–Prandtl law.

The second basic assumption is the *vanishing-viscosity principle*:

Second hypothesis The gradient of the average velocity tends to a well-defined limit as the viscosity vanishes.

According to this vanishing-viscosity principle it is possible to represent the Reynolds number functions $A(Re)$ and $\alpha(Re)$ that enter the relation (11.22) in the form

$$A = A_0 + A_1 \vartheta, \qquad \alpha = \alpha_0 + \alpha_1 \vartheta. \tag{11.23}$$

Here ϑ is a small parameter vanishing at $Re = \infty$, whereas the constants A_0, A_1, α_0 and α_1 should be universal. Thus, the relation (11.22) can be rewritten in the form

$$
\begin{aligned}
\frac{du}{dz} &= \frac{u_*}{z} (A_0 + A_1\vartheta) \left(\frac{u_*z}{\nu}\right)^{\alpha_0 + \alpha_1\vartheta} \\
&= \frac{u_*}{z} (A_0 + A_1\vartheta) \exp\left[(\alpha_0 + \alpha_1\vartheta) \ln \frac{u_*z}{\nu}\right] \\
&= \frac{u_*}{z} (A_0 + A_1\vartheta) \exp\left[\alpha_0 \ln \frac{u_*z}{\nu} + \alpha_1\vartheta \ln \frac{\bar{u}d}{\nu} + \alpha_1\vartheta \ln \frac{z}{d} + \alpha_1\vartheta \ln \frac{u_*}{\bar{u}}\right].
\end{aligned}
$$

(11.24)

We recall that $\bar{u}d/\nu = Re$ is the Reynolds number. When the viscosity ν tends to zero, the condition $\alpha_0 = 0$ is necessary for the existence of a well-defined limit of the velocity gradient (11.24). Therefore, according to the vanishing-viscosity principle, $\alpha_0 = 0$.

Furthermore, remember that the small parameter ϑ is a function of the Reynolds number that must vanish at $Re = \infty$. It will be shown in the next section that the ratio $u_*/\bar{u}$ is of the order of $1/\ln Re$, so that $\ln(u_*/\bar{u})$ is of the order of $\ln \ln Re$ and, at very large Reynolds number, is small in comparison with $\ln Re$. The value of $\ln(z/d)$ is fixed.

Now, a very important step follows. The relation (11.24) demonstrates that if ϑ tends to zero as $Re \to \infty$ faster than $1/\ln Re$ (again recall that $Re = \bar{u}d/\nu$) then the argument of the exponent in (11.24) tends to zero. We have returned to the case of complete similarity, which does not agree with the experimental data. However, if ϑ tends to zero slower than $1/\ln Re$ then a well-defined limit of expression (11.24) for the velocity gradient does not exist, in contrast with the second hypothesis, the vanishing-viscosity principle. Therefore, the only choice compatible with our basic hypotheses is

$$
\vartheta = \frac{1}{\ln Re}.
$$

(11.25)

Here the coefficient of $1/\ln Re$ is taken to equal 1 because if it were not then the constants A_1 and α_1 could simply be renormalized. By integration we obtain

$$
\phi = \frac{u}{u_*} = (C_0 \ln Re + C_1) \left(\frac{u_*z}{\nu}\right)^{\alpha_1/\ln Re}.
$$

(11.26)

Thus, we arrive at a conclusion which is in agreement with the intuitive idea of Prandtl, expressed in his comment about von Kármán's lecture; see Section 11.4.

Indeed, the dimensionless distance from the wall $\eta = u_*z/\nu$ does contain the viscosity and at arbitrary large but finite Reynolds numbers the viscosity influences the velocity gradient. This influence is transmitted, as Prandtl claimed, by wall streaks arriving at the main flow from the viscous sublayer 1 in Figure 11.4, where turbulence and viscosity act together. It is exactly these streaks that create the "intermittency" of wall-bounded flows, i.e. the inhomogeneity in the distribution of turbulent vortices and in the dissipation of turbulent energy into heat. So, *Prandtl's comment is in agreement with the idea of incomplete similarity.*

From (11.26), the parameters u_* and v form a Reynolds-number-dependent dimensional monomial

$$C = (C_0 \ln Re + C_1)u_*^{1+\alpha_1/\ln Re} v^{-\alpha_1/\ln Re} . \tag{11.27}$$

We emphasize that the dimension of this monomial cannot be obtained from dimensional analysis. It determines the power-law velocity distribution in the basic intermediate region 3 of turbulent flow in a pipe, i.e.

$$u = Cz^{\alpha_1/\ln Re} . \tag{11.28}$$

The first comparison of the power law (11.28) with experiment (see the review article by Barenblatt, Chorin and Prostokishin, 1997, and the references therein) was performed on the basis of the experimental data of Nikuradze (1932), obtained under the direct guidance of Prandtl. This comparison confirmed the law (11.28) and made it possible to obtain values of the constants C_0, C_1 and α_1:

$$C_0 = \frac{1}{\sqrt{3}} \approx 0.577\,35 , \qquad C_1 = \frac{5}{2} , \qquad \alpha_1 = \frac{3}{2} . \tag{11.29}$$

In fact, statistical processing has given the following experimental values of the constants C_0 and C_1: $C_0 = 0.578 \pm 0.001$ and $C_1 = 2.50 \pm 0.016$. Accepting the values $C_0 = 1/\sqrt{3}\,(\approx 0.577\,35)$ and $C_1 = 5/2$, we can simplify the calculations while remaining within the limits of accuracy of the experimental data.

Thus, the final result for the velocity distribution in the intermediate region 3 of turbulent flow in pipes is

$$\phi = \frac{u}{u_*} = \left(\frac{1}{\sqrt{3}} \ln Re + \frac{5}{2}\right) \eta^{\frac{3}{2\ln Re}} , \tag{11.30}$$

where $\eta = u_*z/v$, or, equivalently,

$$\phi = \left(\frac{\sqrt{3} + 5\alpha}{2\alpha}\right) \eta^\alpha , \qquad \alpha = \frac{3}{2 \ln Re} . \tag{11.31}$$

An instructive form of (11.31) can be obtained by introducing a new variable ψ:

$$\psi = \frac{1}{\alpha} \ln \frac{2\alpha\phi}{\sqrt{3} + 5\alpha} . \tag{11.32}$$

In this case relation (11.31) is reduced to the simple form

$$\psi = \ln \eta . \tag{11.33}$$

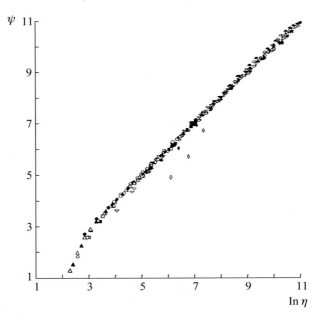

Figure 11.7 Experimental data points of Nikuradze (1932) in $(\ln \eta, \psi)$ coordinates at $\eta > 30$ lie close to the bisectrix of the first quadrant, confirming the scaling law (11.30), (11.31). From Barenblatt, Chorin and Prostokishin (1997). The values of *Re* are as follows: $\triangle$, 4×10^3; $\blacktriangle$, 6.1×10^3; $\circ$, 9.2×10^3; $\bullet$, 1.66×10^4; $\square$, 2.33×10^4; $\blacksquare$, 4.34×10^4; ∇, 1.05×10^5; $\blacktriangledown$, 2.05×10^5; $\triangledown$, 3.96×10^5; $\blacktriangledown$, 7.25×10^5; $\diamond$, 1.11×10^6; $\blacklozenge$, 1.536×10^6; $+$, 1.959×10^6; $\times$, 2.35×10^6; $\frown$, 2.79×10^6; $\blacktriangle$, 3.24×10^6.

Equation (11.33) is particularly important. As was noted previously, a key difference between the universal, Reynolds-number-independent, logarithmic law obtained on the basis of the assumption of complete similarity and the Reynolds-number-dependent scaling law (11.30), (11.31) is as follows. While the former implies that all the data points in the $(\ln \eta, \phi)$-plane cluster on a single curve, the latter predicts a separate curve $\phi(\ln \eta)$ for each Reynolds number, so that the corresponding data points should fill out an area in the $(\ln \eta, \phi)$-plane. The transformation (11.32) reverses the roles of the two laws. According to the Reynolds-number-dependent scaling law (11.30), (11.31) all the data points should cluster in the $(\ln \eta, \psi)$-plane on a single curve, and a particularly simple one at that, the bisectrix of the first quadrant. By contrast, if the universal law (11.20) holds, the data points should be area-filling.

Figure 11.7 is instructive. All Nikuradze's (1932) experimental data (256 points) are presented in the $(\ln \eta, \psi)$-plane. We observe that for $\eta > 25$ (i.e. outside the viscous sublayer), all the data, except for a very few, fall on the bisectrix, confirming the validity of the scaling law. A plausible explanation for the four exceptional points that lie far from the bisectrix, which has been offered by Professor D. Coles, is that they are typographical errors. However, we shall not presume to make corrections to the data published by Nikuradze (1932). It is also important to mention that overall Figure 11.7 testifies to the self-consistency of Nikuradze's data.

11.6 Reynolds-number dependence of the drag in pipes following from the power law

It is instructive to compare the prediction of the Reynolds-number dependence of the drag coefficient in pipes that follows from the scaling law (11.30), (11.31) with the experimental data. Also instructive is a comparison of the experimental data with the Reynolds-number dependence of the drag coefficient that follows from the von Kármán–Prandtl law.

The standard definition of the drag coefficient λ in the engineering literature is

$$\lambda = \frac{\tau}{\rho \bar{u}^2 / 8} = 8 \left(\frac{u_*}{\bar{u}} \right)^2 . \tag{11.34}$$

The average velocity (the total discharge volume per unit time per unit cross-sectional area) is given by

$$\bar{u} = \frac{8}{d^2} \int_0^{d/2} u(z) \left(\frac{d}{2} - z \right) dz . \tag{11.35}$$

To calculate $\bar{u}$ we assume that it is possible to replace $u(z)$ in (11.35) by the scaling law (11.31), which is strictly true only in the intermediate region 3 (see Figure 11.4). This replacement introduces a systematic error, which, however, is small at very large Reynolds numbers, when the contributions from the near-wall region 1 and the near-axis region 2 are small. The result is

$$\bar{u} = u_* \frac{\sqrt{3} + 5\alpha}{\alpha} \left(\frac{u_* d}{\nu} \right)^\alpha \frac{1}{2^\alpha (1 + \alpha)(2 + \alpha)} . \tag{11.36}$$

However, $\alpha = 3/(2 \ln Re)$, so that $Re = \exp[3/(2\alpha)]$, and therefore

$$\frac{u_* d}{\nu} = \frac{e^{3/(2\alpha)} 2^\alpha \alpha (1 + \alpha)(2 + \alpha)}{\sqrt{3} + 5\alpha} . \tag{11.37}$$

Some further simple algebra yields an explicit relationship between the dimensionless drag coefficient λ and the Reynolds number Re:

$$\lambda = \frac{8}{F^{2(1+\alpha)}} , \tag{11.38}$$

where

$$F = \frac{e^{3/2}(\sqrt{3} + 5\alpha)}{2^\alpha \alpha (1 + \alpha)(2 + \alpha)} , \qquad \alpha = \frac{3}{2 \ln Re} .$$

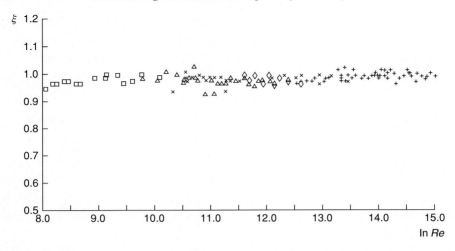

Figure 11.8 Nikuradze's (1932) data for various pipes and various Reynolds numbers confirm to high accuracy the friction law (11.50), (11.51) which follows from the scaling law (11.30), (11.31); the figure shows $\xi = \lambda_{\exp}/\lambda_{\text{pred}}$ for various pipe radii: □, $d = 1$ cm; △, $d = 2$ cm; ◊, $d = 3$ cm; ×, $d = 5$ cm; +, $d = 10$ cm. From Barenblatt, Chorin and Prostokishin (1997).

In the same paper by Nikuradze (1932) there were presented (also in digital form) data from drag measurements, independent of the velocity measurements. The values of the ratio $\xi = \lambda_{\exp}/\lambda_{\text{pred}}$ as a function of Re for all 125 data points presented in the Nikuradze (1932) paper are shown in Figure 11.8. Here $\lambda_{\exp}$ is the experimental value presented in Nikuradze's paper and λ_{pred} is the value given by the relationship (11.38), which follows from the scaling law (11.31). In the ideal case ξ will be equal to 1. In fact, in the figure it is not quite equal to 1 but the difference is within the bounds of experimental error, although the slight systematic deviation could be ascribed to the fact, mentioned above, that the scaling law (11.31) is not valid in the viscous sublayer and near the axis.

In contrast with the explicit formula (11.38), the formula obtained by Prandtl for the drag coefficient λ following from the universal logarithmic law is an implicit one. For $\kappa = 0.4$ and $C = 5.5$ it is as follows:

$$\frac{1}{\sqrt{\lambda}} = 2.035 \log(\sqrt{\lambda}\, Re) - 0.91 . \tag{11.39}$$

Schlichting (1968) proposed a *corrected Prandtl's drag law*, also an implicit one, obtained by adjusting the constants in (11.39) for better correspondence with the experimental data:

$$\frac{1}{\sqrt{\lambda}} = 2.0 \log(\sqrt{\lambda}\, Re) - 0.8 . \tag{11.40}$$

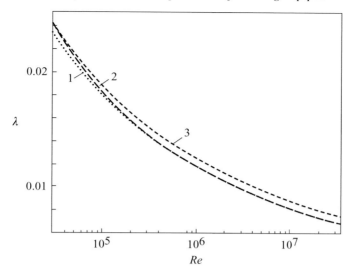

Figure 11.9 The drag coefficient (friction parameter) λ as a function of ln Re, obtained as follows: curve 1, λ as given by (11.38) as a consequence of the power law; curve 2, λ as given by the "corrected" Prandtl law (11.40) designed to fit the data; curve 3, λ as derived from the logarithmic law using the constants $\kappa = 0.44$, $C = 6.3$ suggested by the Princeton group. From Barenblatt, Chorin and Prostokishin (1997).

So, summarizing, Figure 11.9 displays curves corresponding to three expressions for the drag coefficient $\lambda\,(Re)$:

curve 1 is given by the relation (11.38) following from the scaling law (11.30), (11.31);

curve 2 is given by the corrected Prandtl law (11.40) designed to fit the data;

curve 3 is derived from the universal logarithmic law using the constants $\kappa = 0.44$, $C = 6.3$ suggested by Zagarola (1996) (see Figure 11.6).

The agreement between curves 1 and 2 speaks for itself. We emphasize that no further manipulation of the constants entering the power law or the relation (11.38) is needed to bring this relation in line with the experimental data.

We mention in conclusion that from (11.36) and the relation $Re = \exp[3/(2\alpha)]$ the following relation for $u_*/\bar{u}$ is obtained:

$$\frac{u_*}{\bar{u}} = \frac{e^{3/[2(1+\alpha)]}[\alpha 2^\alpha(1+\alpha)(2+\alpha)]^{1/(1+\alpha)}}{(\sqrt{3}+5\alpha)^{1/(1+\alpha)}}, \qquad \alpha = \frac{3}{2\ln Re}. \qquad (11.41)$$

At very large *Re*, when $\alpha = 3/(2 \ln Re)$ is small, the asymptotic relation comes from (11.41):

$$\frac{u_*}{\bar{u}} = \frac{\sqrt{3}e^{3/2}}{\ln Re} ,$$

(11.42)

so that $\ln(u_*/\bar{u})$ is equal to $- \ln \ln Re + O(1)$. This justifies the neglect of the term $\alpha_1 \vartheta \ln(u_*/\bar{u})$ in comparison with the term $\alpha_1 \vartheta \ln Re$ in (11.24), at very large *Re*.

11.7 Further comparison of the Reynolds-number-dependent scaling law and the universal logarithmic law

The scaling law (11.31) can be represented in the form

$$\phi = \left(\frac{1}{\sqrt{3}} \ln Re + \frac{5}{2}\right) \exp\left(\frac{3 \ln \eta}{2 \ln Re}\right) .$$

(11.43)

This form clearly reveals the self-similarity property of the scaling law curves in the $(\ln \eta, \phi)$-plane: the curves for different *Re* can be obtained from one another by a similarity transformation: in the "reduced" variables

$$X = \frac{3 \ln \eta}{2 \ln Re} , \qquad Y = \phi\left(\frac{1}{\sqrt{3}} \ln Re + \frac{5}{2}\right)^{-1}$$

(11.44)

all the curves of the scaling law collapse onto a single curve. For each value of *Re* in the $(\ln \eta, \phi)$-plane a distinct curve $\phi(\ln \eta)$ is obtained. This is in contrast with the prediction of the universal logarithmic law. According to this law, in the $(\ln \eta, \phi)$-plane all the data points should lie on a single curve,

$$\phi = \frac{1}{\kappa} \ln \eta + C .$$

(11.45)

The family of curves (11.43) has an envelope. It is obtained by eliminating *Re* – the parameter of the family – between equation (11.43) and the equation $\partial_{\ln Re}\phi = 0$. The latter equation can be written as

$$\frac{3 \ln \eta}{2 \ln Re} = \frac{\sqrt{3} \ln \eta}{10}\left[\left(1 + \frac{20}{\sqrt{3} \ln \eta}\right)^{1/2} - 1\right] .$$

(11.46)

The envelope is shown in Figure 11.10 with the straight line $\phi = 2.5 \ln \eta + 5.5$ taken by Schlichting (1968) as the representation of the universal logarithmic law. It is clear that in the range of $\ln \eta$ under consideration these curves are close and nearly parallel. Moreover, if the constant 5.5 is replaced by 5.1, the two curves practically coincide. The line $\phi = 2.5 \ln \eta + 5.1$, however, is the well-known representation of the universal logarithmic law given, for instance, in the book by Monin and Yaglom (1971)!

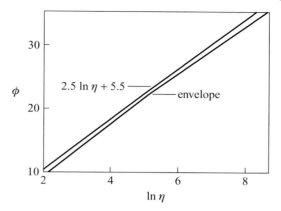

Figure 11.10 The envelope of the scaling-law curves in the $(\ln \eta, \phi)$-plane is very close to the generally accepted straight line of the universal logarithmic law, even at moderate η. From Barenblatt, Chorin and Prostokishin (1997).

Now, we take the following step. If one allows $\ln Re$ and $\ln \eta$ to tend to ∞ while one remains on the envelope then the right-hand side of equation (11.46) tends to 1, so the equation of the envelope must take the form

$$\phi = \frac{\sqrt{3}}{2} e \ln \eta + \frac{5}{2} e + \text{small quantities} . \qquad (11.47)$$

The value $2/\sqrt{3} e = 0.425 \ldots$ is close to the value of the von Kármán constant $\kappa = 0.417$ obtained by Nikuradze (1932). However, the value of the additive constant $5e/2 \simeq 6.79$ is substantially higher than the values commonly ascribed to the additive constant $C = 5.1$ or 5.5 in the universal logarithmic law; the value 6.3, suggested by Zagarola (1996), is a rare exception. The value 6.79 is closer to the value of the additive constant of the envelope of the scaling-law curves. The reason is that for this asymptotic value to be observed the values of $\ln \eta$ and $\ln Re$ have to be large enough for two things to happen simultaneously: the asymptotic regime must be reached on the envelope while the envelope still approximates the individual curves of the family corresponding to the scaling law. However, velocity profiles satisfying such conditions have practically never been measured in experiments with pipe flow. (The experiments of Zagarola (1996) are an important exception.)

Figure 11.11 presents three of Nikuradze's experimental runs, with Reynolds numbers differing by approximately an order of magnitude, namely, 1.67×10^4, 2.05×10^5 and 3.24×10^6. The corresponding scaling-law curves, i.e. the straight line of the universal logarithmic law and the envelope of the family of scaling-law curves, are also exhibited in Figure 11.11. It can be seen that the scaling-law curves

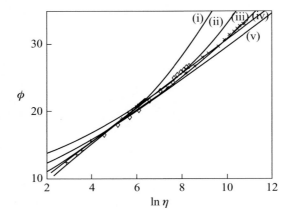

Figure 11.11 Experimental points and the scaling law curves at various Reynolds numbers: (i) +, $Re = 1.67 \times 10^4$; (ii) ◊, $Re = 2.05 \times 10^5$; (iii) ×, $Re = 3.24 \times 10^6$; (iv) the universal logarithmic law; and (v) the envelope. From Barenblatt, Chorin and Prostokishin (1997).

have a small systematic and discernible advantage over the universal logarithmic law. However, at this stage we do not wish to emphasize this advantage. We would like to point out here that Nikuradze's experimental data all correspond to points near the envelope of the scaling law curves! Later we shall present far more decisive criteria for deciding between the Reynolds-number-dependent scaling law and the universal logarithmic law.

Having fitted constants to the scaling law from the available data, which are close to the envelope, we shall now extrapolate the resulting law to points farther from the envelope. We emphasize that the constants are universal and their values (11.29) are fixed. Therefore, they should be the same for all data points from all experiments whose quality matches that of Nikuradze's (1932) experiments. If this extrapolation were successful in predicting the data, the result would be a dramatic validation of the scaling law. The extrapolation will be carried out with the help of vanishing-viscosity asymptotics, based on the second hypothesis formulated at the start of Section 11.5.

Consider again the scaling law (11.30), (11.31). At a fixed distance from the wall, in a specific pipe with a given pressure gradient, one is not free to vary $Re = \bar{u}d/\nu$ and $\eta = u_*z/\nu$ independently because the viscosity ν enters both parameters. If ν is decreased by an experimenter then the two quantities Re and η will increase in a self-consistent way. The limit of the velocity gradient that corresponds to the experimental situation is the limit of vanishing viscosity. The existence of such a limit is asserted by the above-mentioned second hypothesis. So, when one takes the limit of vanishing viscosity, one is considering flows at ever larger η and ever larger Re. The ratio $3 \ln \eta / (2 \ln Re)$ tends to 3/2 because ν appears in the same way in both numerator and denominator.

We show this now in more detail. Consider the combination $3 \ln \eta / (2 \ln Re)$. It can be represented in the form

$$\frac{3 \ln \eta}{2 \ln Re} = \frac{3[\ln(\bar{u}d/\nu) + \ln(u_*/\bar{u}) + \ln(z/d)]}{2 \ln(\bar{u}d/\nu)}. \tag{11.48}$$

It was shown in the previous section, in equation (11.42), that the term $\ln(\bar{u}/u_*)$ is of order $\ln \ln Re$ at very large Reynolds numbers $Re = \bar{u}d/\nu$, which is asymptotically small in comparison with the term $\ln Re$ and so can be neglected. The crucial point is that owing to the small value of the viscosity the term $\ln Re = \ln(\bar{u}d/\nu)$ is also dominant in the numerator as long as the ratio z/d remains bounded from below by, for example, a predetermined fraction. Thus, as long as one stays away from a suitable neighborhood of the wall, the ratio $3 \ln \eta / (2 \ln Re)$ is close to 3/2 (z is obviously bounded by $d/2$). Therefore the quantity $1 - \ln \eta / \ln Re$ can be considered to be a small parameter as long as $z > \Delta$, where Δ is a predetermined fraction of the diameter d. Then the quantity $\exp[3 \ln \eta / (2 \ln Re)]$ is approximately equal to

$$\exp\left[\frac{3}{2} - \frac{3}{2}\left(1 - \frac{\ln \eta}{\ln Re}\right)\right] \approx e^{3/2}\left[1 - \frac{3}{2}\left(1 - \frac{\ln \eta}{\ln Re}\right)\right] = e^{3/2}\left[\frac{3}{2}\frac{\ln \eta}{\ln Re} - \frac{1}{2}\right]. \tag{11.49}$$

According to (11.30) we also have

$$\eta \partial_\eta \phi = \partial_{\ln \eta} \phi = \left(\frac{\sqrt{3}}{2} + \frac{15}{4 \ln Re}\right) \exp\left(\frac{3 \ln \eta}{2 \ln Re}\right), \tag{11.50}$$

and the approximation (11.49) can also be used in (11.50). Thus, in the intermediate asymptotic range of distances z such that $z > \Delta$ but at the same time less than $d/2$, the following asymptotic relations for the scaling law should hold as $Re \to \infty$:

$$\phi = e^{3/2}\left(\frac{\sqrt{3}}{2} + \frac{15}{4 \ln Re}\right) \ln \eta - \frac{e^{3/2}}{2\sqrt{3}} \ln Re - \frac{5}{4} e^{3/2} \tag{11.51}$$

and

$$\partial_{\ln \eta} \phi = e^{3/2}\left(\frac{\sqrt{3}}{2} + \frac{15}{4 \ln Re}\right). \tag{11.52}$$

At the same time, formula (11.47) shows that for the envelope of the power-law curves the asymptotic relation is

$$\partial_{\ln \eta} \phi = \frac{\sqrt{3}}{2} e. \tag{11.53}$$

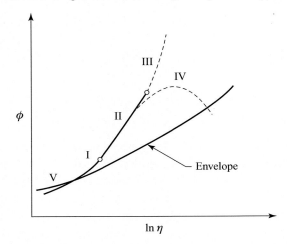

Figure 11.12 The individual members of the family of scaling laws (11.30), (11.31) near the envelope in the $(\ln \eta, \phi)$-plane have a straight intermediate interval with a slope substantially larger than that of the envelope: I, the part close to the envelope; II, the straight intermediate part; III, the fast-growing ultimate part having no physical meaning because there are no corresponding points in the pipe; IV, the region near the axis of the pipe where the scaling law is invalid; V, the part which was never observed because of the large diameter of the gauge. From Barenblatt, Chorin and Prostokishin (1997).

The ratio, $e^{1/2}[1 + 15/(2\sqrt{3} \ln Re)]$, of the slopes (11.52) and (11.53) is significant. It shows that individual members of the family (11.30) at large Re should have an intermediate part represented in the $(\ln \eta, \phi)$-plane by straight lines, with a slope different from the slope of the envelope by a factor larger than $\sqrt{e} \simeq 1.65$. Therefore the graphs of the individual members of the family (11.30) should have the form presented schematically in Figure 11.12.

11.8 Modification of the Izakson–Millikan–von Mises analysis of the flow in the intermediate region

We now examine in detail a well-known argument determining the structure of the flow in the intermediate region, due to Izakson (1937), Millikan (1939) and von Mises (1941) (IMM). In this argument it is assumed that from the wall outward, for some distance, a generalized *law of the wall* is valid:

$$\phi = u/u_* = f(u_* z/\nu) \,, \tag{11.54}$$

where f is a dimensionless function. The influence of the Reynolds number containing the external length scale is neglected (for pipe flow this is the diameter d of the pipe). Heuristically, acceptance of the law (11.54) means that it is assumed that the fluid flow near enough to a smooth wall does not feel the outer part of the flow. However, at points adjacent to the axis of the pipe in the pipe flow one assumes a *defect law*,

$$u_{CL} - u = u_* g(2z/d) , \tag{11.55}$$

where u_{CL} is the average velocity at the centerline and g is another dimensionless function. Here the neglect of the effect of Re means the neglect of viscosity effects. In other words this means that near the axis, where the *averaged* velocity gradients are small, one is assuming that the viscosity is unimportant.

Taken together, these two assumptions constitute an assumption of the *separation of scales*, according to which, at large enough but finite values of Re, viscous scales and inviscid scales can be studied in partial isolation. Self-consistency then demands that for some interval in z the laws (11.54) and (11.55) overlap, so that

$$u_{CL} - u = u_{CL} - f(u_* z/\nu) = u_* g(2z/d) . \tag{11.56}$$

After differentiating (11.56) with respect to z and then multiplying by z one obtains

$$\eta f'(\eta) = -\xi g'(\xi) = \frac{1}{\kappa} , \tag{11.57}$$

where $\eta = u_* z/\nu$, $\xi = 2z/d$ and κ is a constant. Integration then yields the universal, Reynolds-number-independent, law of the wall,

$$f(\eta) = \frac{1}{\kappa} \ln \eta + B , \tag{11.58}$$

if the additive constant is assumed to be Re-independent, as well as the defect law

$$g(\xi) = -\frac{1}{\kappa} \ln \xi + B_* , \tag{11.59}$$

with Reynolds-number-dependent additive constant

$$B_* = \frac{u_{CL}}{u_*} - \frac{1}{\kappa} \ln Re - \frac{1}{\kappa} \ln \frac{u_*}{\bar{u}} - B . \tag{11.60}$$

However, there is an obvious contradiction: owing to the matching with the universal logarithmic law, the defect law appears to be Re-dependent. So, our previous analysis and the experimental data (see Figure 11.6) suggest a different approach, that the original IMM argument is elegant but oversimplified. We shall see that, when suitably improved, the IMM argument survives and supports our conclusions.

We begin by noting that in the nearly linear part, II, of the graph in Figure 11.12 the flow can be described by a local logarithmic law with a Reynolds-number-dependent effective von Kármán constant $\kappa_{\text{eff}} = \kappa(Re)$:

$$k_{\text{eff}} = e^{-3/2} \left(\frac{\sqrt{3}}{2} + \frac{15}{4 \ln Re} \right)^{-1} . \tag{11.61}$$

As $Re \to \infty$, κ_{eff} tends, although very slowly, to the limit $\kappa_\infty = 2/(\sqrt{3}\, e^{3/2}) \sim 0.2776$, which is smaller than the commonly accepted von Kármán constant by a factor of nearly two. For finite although very large Re, κ_{eff} is even smaller. With this in mind, the IMM procedure can be modified as follows. The law of the wall becomes

$$\phi = u/u_* = f(u_* z/\nu, Re)\,, \tag{11.62}$$

where the form of f is given by the expression (11.51). The defect law (11.55) becomes

$$u_{\mathrm{CL}} - u = u_* g(2z/d, Re)\,. \tag{11.63}$$

Now we assume that the laws (11.62) and (11.63) overlap on some z interval and obtain

$$g(2y/d, Re) = \phi_{\mathrm{CL}} - \left(\frac{1}{\sqrt{3}}\ln Re + \frac{5}{2}\right)e^{3/2} - \left(\frac{\sqrt{3}}{2} + \frac{15}{4\ln Re}\right)\ln\frac{u_*}{\bar{u}}\,,$$

$$\phi_{\mathrm{CL}} = \frac{u_{\mathrm{CL}}}{u_*}\,. \tag{11.64}$$

We remember that $u_*/\bar{u}$ is a known function of the Reynolds number, given by the relation (11.41). This calculation is self-consistent and differs from the original IMM procedure: a Reynolds-number-dependent defect law is matched to the actual curves of the scaling law rather than to their envelope, *misinterpreted as being identical to the actual curves*. The modified matching was successfully carried out because the scaling law has an intermediate range that is approximately linear in $\ln \eta$ (see Figure 11.6). The success of the matching does not depend on the specific values of the constants C_0, C_1 and α_1 in (11.26).

Note also that the inner and outer portions of the flow "feel" each other for all finite values of Re. This coupling disappears only in the limit of vanishing viscosity. Note further that as the viscosity is decreased beyond the point where the chevron kink (a sharp deviation from the envelope) appears, the location of this kink moves slowly towards the wall of the pipe. At extremely high Reynolds numbers, well beyond those currently achievable, the power law collapses onto the upper part of the chevron, resulting in an apparently new logarithmic law with constants different from the usually accepted constants in the von Kármán–Prandtl law. In particular the new value of the von Kármán constant would be, as was shown above, $\kappa_\infty = 2/(\sqrt{3}\, e^{3/2}) \approx 0.2776$. It is important to note that this new logarithmic law, corresponding to the upper part of the chevron, does not lie in the region where the usual von Kármán–Prandtl law applies.

11.9 Further comparison of scaling laws with experimental data

The dramatic feature of the velocity profile in the $(\ln \eta, \phi)$-plane at small viscosity ν, predicted in the previous section (see Figure 11.12), is its chevron form: the limits as $\nu \to 0$ of the scaling-law curves have a kink where they leave the envelope, and the difference in the slopes of the two branches of the chevron is substantial, more than $\sqrt{e} \sim 1.65$.

Many experimental confirmations of this behavior exist, in both old and new experiments. In the next section we will consider the results of experiments on boundary layer flows. Here we will discuss the experimental study of pipe flows by the Princeton group (Zagarola, 1996; Zagarola *et al.*, 1996) in comparison with the results in Nikuradze (1932). Zagarola's publications contain many new data points for pipe flow obtained in the high-pressure pipe flow of air (using a "Superpipe"); this idea was proposed by the remarkable experimentalist Professor G. Brown. A high pressure increases the density ρ and also increases the dynamic viscosity but at a much smaller rate; the kinematic viscosity is decreased substantially and thus the Reynolds number Re is increased. It was claimed that in this way one can increase the Reynolds number by an order of magnitude over the Re values achieved by Nikuradze with a flow of water.

It will be shown later that for Reynolds numbers $Re \gtrsim 10^6$ the Princeton data contain a systematic error growing with Re. Nevertheless, as one can clearly see from Figure 11.6, the appearance of a chevron structure and the splitting of the velocity curves according to their Reynolds number are so marked that even a systematic error at large Re cannot destroy them. To each Re value there corresponds a separate curve in the $(\ln \eta, \phi)$-plane, with a pronounced linear part whose slope is larger than the slope of the envelope by a factor that is always larger than 1.5. For smaller values of z, the deviation of the curves from their envelope is small (corresponding to part I of the curve in Figure 11.12).

We consider the graph in Figure 11.6 to be a clear confirmation of the proposed scaling law and a strong argument against the universal logarithmic law.

We emphasize the following important consequences of the vanishing-viscosity analysis of the scaling-law curves and of the experimental data.

(1) Both linear segments of the piecewise linear chevron structure have the same scaling law: the constants that describe the inner segment (i.e. the segment closer to the wall) also describe the outer segment.

(2) The overlap (central) region is the outer segment of the chevron; thus the outer segment belongs both to the wall region and to the defect region.

(3) There is no other possible locus for the overlap; since the slopes of the outer and inner segments tend to two different constants as $\nu \to 0$, the segments can never overlap on the inner segment.

(4) The whole chevron constitutes a single law; the possibility that the inner segment is described by the universal (i.e. *Re*-independent) logarithmic law is excluded.

(5) More generally, since the defect law (11.63) must be a concave-downward function of z/d (the second derivative is negative), the only way in which there can ever be a portion of the velocity profile that is concave upwards outside the near vicinity of the wall, as observed in Figure 11.6, is to let the overlap region be concave upward, as we are proposing, rather than straight.

We note that the prediction of a difference larger than $\sqrt{e}$ between the slopes of the individual velocity profiles and the slope of their envelope provides an easily verified criterion for assessing the agreement between the experimental data and the scaling law. *At high Re the difference between the proposed scaling law and the universal logarithmic law is large enough to have a substantial impact on the outcome of engineering calculations.*

Now we come to a more detailed comparison of the proposed scaling law with the data presented by the Princeton group. The advantage of this set of data for such comparisons is that, like Nikuradze's data, they are presented in tabular form. The Princeton group presented the results of 26 experimental runs, each run containing data from measurements of the velocity distribution over the cross-section of the pipe as well as the measured drag coefficients. The experiments were performed with air flow in a pipe at high pressure (the pressure varied from ~ 1 to ~ 190 atmospheres). The kinematic viscosity of air under normal conditions is $\sim 0.15\ \text{cm}^2/\text{s}$ and that of water is $\sim 0.01\ \text{cm}^2/\text{s}$; therefore the Princeton group had to compress the air to roughly 15 atmospheres to reach the kinematic viscosity of water. As we will see later, it is in this respect that the Princeton experiments became inaccurate.

Another important advantage of the Princeton data is that they contain many experimental points far from the envelope (see Figure 11.6). In the published experiments of Nikuradze there were no such data. Therefore the most interesting step is the comparison of the Princeton data with the scaling law (11.30), (11.31) according to the same procedure as in Section 11.5. Thus all the Princeton data were plotted in the $(\ln \eta, \psi)$-plane, where, as before,

$$\psi = \frac{1}{\alpha} \ln \frac{2\alpha\phi}{\sqrt{3}+5\alpha}, \qquad \alpha = \frac{3}{2 \ln Re}, \qquad Re = \frac{\bar{u}d}{\nu}, \qquad \phi = \frac{u}{u_*}. \qquad (11.65)$$

For the first ten runs, *Re* was as follows:

$$3.16 \times 10^4, \qquad 4.17 \times 10^4, \qquad 5.67 \times 10^4, \qquad 7.43 \times 10^4, \qquad 9.88 \times 10^4,$$
$$1.46 \times 10^5, \qquad 1.85 \times 10^5, \qquad 2.30 \times 10^5, \qquad 3.09 \times 10^5, \qquad 4.09 \times 10^5.$$

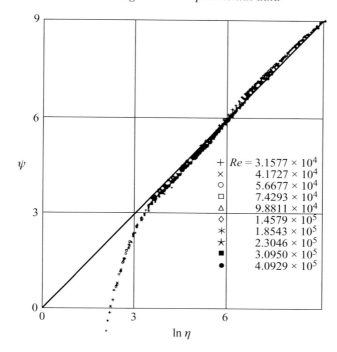

Figure 11.13 The lower-Reynolds-number Princeton data are in agreement with the scaling law: in the $(\ln \eta, \psi)$-plane they are close to the bisectrix (except, as expected, in the near-axis region). From Barenblatt, Chorin and Prostokishin (1997).

The data are presented in Figure 11.13. It is seen that, as in the case of Nikuradze's data, the experimental points after $\eta = 25$ are concentrated near the bisectrix of the first quadrant, as expected according to the model presented above. The points close to the pipe axis should be removed because the scaling law is invalid for them. It was in fact sufficient to remove only the points where $2z/d$ was more than 0.95.

However, for the last six runs, where Re was

$$1.02 \times 10^7, \qquad 1.36 \times 10^7, \qquad 1.82 \times 10^7,$$
$$2.40 \times 10^7, \qquad 2.99 \times 10^7, \qquad 3.52 \times 10^7,$$

the situation is different: the experimental points for these runs are concentrated (for $z/R < 0.95$, $R = d/2$) along straight lines parallel to the bisectrix but not on the bisectrix itself (Figure 11.14). Note that all these curves present a chevron and that there is a separate curve for each value of Re; the advantage of the scaling law over the universal logarithmic law is not in question even in the presence of this disturbing shift in the processed curves.

Some hint to what happens was given by a comparison of the experiments of Nikuradze and those of the Princeton group performed at roughly equal Reynolds

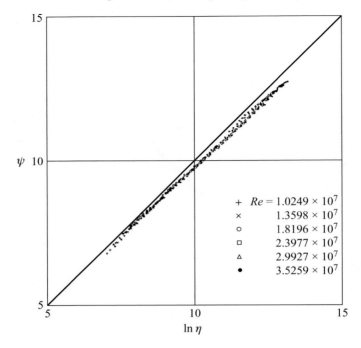

Figure 11.14 There is a noticeable disagreement between the large Reynolds numbers Princeton data and the prediction of the scaling law (11.31). In the $(\ln \eta, \psi)$-plane they concentrate along lines parallel to the bisectrix, not on the bisectrix itself. The points with $2z/d > 0.95$ are excluded; they are in the near-axis region. From Barenblatt, Chorin and Prostokishin (1997).

numbers. There are six such experiments and for five of them, at moderate Reynolds numbers, a satisfactory coincidence was found. This coincidence means that the scaling law (11.30), (11.31) is also confirmed by the Princeton experiments. However, for the Princeton run #16, with $Re = 2.345 \times 10^6$, and the corresponding run of Nikuradze, with $Re = 2.35 \times 10^6$, which, like the other Nikuradze runs, corresponds quite satisfactorily to the scaling law, a noticeable disagreement was found (Figure 11.15). In the main part of the $(\ln \eta, \psi)$-plane there is a nearly uniform shift along the $\ln \eta$ axis. What can be the meaning of such a shift? If both u_* and z were measured correctly, the most likely source of the discrepancy is in the determination of the viscosity. It is of importance that the pressure gradients in these experiments were small enough not to create a variation of the viscosity along the pipe, and thus in each run the viscosity can be viewed as a constant.

It was concluded that something had happened in the high-Reynolds-number high-pressure Princeton experiments to move the viscosity that determines the

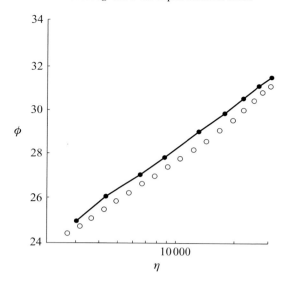

Figure 11.15   The Princeton data at $Re = 2.345 \times 10^6$ (o) and Nikuradze's data
(•) at $Re = 2.35 \times 10^6$. The disagreement is clear. From Barenblatt, Chorin and
Prostokishin (1997).

velocity profile from its actual value to a *shifted* value ν', so that

$$\ln \eta = \ln \frac{u_* y}{\nu} = \ln \frac{u_* y}{\nu'} + \ln \frac{\nu'}{\nu}, \qquad (11.66)$$

and the shift factor $\ln(\nu'/\nu)$ is constant for each run.

To check this conclusion, the following procedure was used for the last six
Princeton runs. For every experimental point of each run, the values of the
difference

$$\chi = \ln \eta - \psi \qquad (11.67)$$

and of $\bar{\chi}$, the mean value of χ per run, were calculated. The dispersion of the quantity was also calculated and found to be very small. Then every experimental point
was shifted by $\bar{\chi}$ inwards along the $\ln \eta$ axis. The results for the unshifted points
are presented in Figure 11.14 and for the shifted points in Figure 11.16. They show
that there exists a single factor per run by which the viscosity is altered and shifts
the velocity profiles at high Reynolds numbers; this does not happen at moderate
Reynolds numbers.

Three possible reasons were investigated.

(1) *Incorrect pressure or temperature measurement.* The density and viscosity
    were not measured directly but were calculated by the Princeton group on
    the basis of the measured pressures and temperatures. Therefore an incorrect

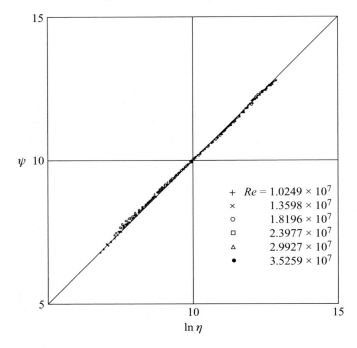

Figure 11.16 After the viscosity correction (constant for each run), the large-Reynolds-number Princeton data agree with the prediction of the scaling law in the $(\ln \eta, \psi)$-plane; the points are close to the bisectrix (except for the near-axis points). From Barenblatt, Chorin and Prostokishin (1997).

pressure measurement could be the reason for the shift discussed above (the measurement of the temperature was not in doubt). After an inspection of the information presented in the thesis of Dr Zagarola (Zagarola, 1996), the conclusion was achieved that this was unlikely.

(2) *Incorrect density and viscosity calculations.* Indeed, the Princeton group used rather old pressure–density relations for their calculations. However, the data of Dr Friend (from the National Institute of Standards and Technology) confirmed the Princeton group's calculations very accurately.

This leaves only one possible explanation for the observed shift in the viscosity. As is well known, if the walls of the pipe are not sufficiently smooth then regions of roughness protrude beyond the viscous sublayer, and a shift in the velocity profile will be observed in the intermediate region, exactly as if the viscosity of the fluid were changed. There is a well-known formula for the equivalent viscosity (see e.g. Monin and Yaglom (1971), p. 286, formula (5.25b)). Therefore, the final possible reason for the shift is as follows.

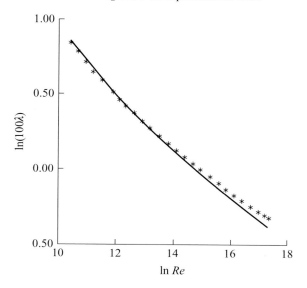

Figure 11.17 The drag coefficient λ as a function of the Reynolds number for the Princeton data; the stars give the Princeton data and the solid line is the law for smooth pipes. From Barenblatt, Chorin and Prostokishin (1997).

(3) *Some roughness of the pipe walls is revealed at large Reynolds numbers.* To check this possibility we turn to well-known data concerning the Reynolds-number dependence of the drag coefficient for flows in rough pipes. The general situation is as follows. For a given mean height of the surface roughness, the data for smooth and rough pipes coincide up to a critical Reynolds number. When this is reached, the Reynolds-number dependence of the drag coefficient for rough pipes deviates from that for smooth pipes. Clearly the critical Reynolds number depends on the mean height of the roughness: the smaller this height, the later the deviation begins.

The drag coefficient as a function of the Reynolds number for the Princeton experiments is presented in Figure 11.17; the solid line corresponds to the theoretical relation (11.38). The graph shows that the deviation starts at approximately $Re = 10^6$. This is a sensitive indicator of the quality of the the velocity profiles; it shows that starting from the run for $Re = 1.02 \times 10^6$ the profiles presented by the Princeton group are inappropriate for comparison with the theoretical predictions for smooth pipes; this is the reason for the observed differences between the predicted profiles and those measured by the Princeton group. This conclusion found further confirmation in the paper by Perry *et al.* (2001), in which the Princeton data were processed using the formulae for rough pipes.

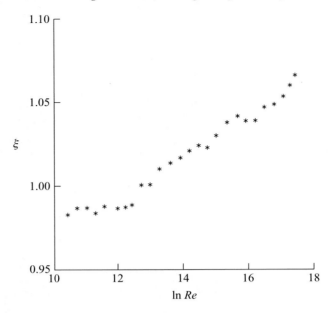

Figure 11.18 The relative friction coefficient $\xi = \lambda_{\text{exp}}/\lambda_{\text{pred}}$ for the Princeton data as a function of ln *Re*. From Barenblatt, Chorin and Prostokishin (1997).

An even sharper visualization of the effect of roughness on the Princeton data is offered in Figure 11.18, where the relative friction coefficient $\xi = \lambda_{\text{exp}}/\lambda_{\text{pred}}$ (already used in Figure 11.8), calculated from these data is plotted as a function of ln *Re*. One can see that starting from approximately $Re = 10^6$ (ln $Re \sim 13.8$) the values of ξ begin to grow steeply (compare with Figure 11.8, where nothing in particular happens at this value of *Re*). If the same viscosity correction as that used in Figure 11.16 is introduced into the calculation of λ_{pred} then the kink disappears, as one can see in Figure 11.19, where $\xi = \lambda_{\text{exp}}/\lambda_{\text{corrected}}$ is plotted. This shows that starting at $Re = 10^6$ a drag estimate based on the assumption that the pipe is smooth becomes increasingly inadequate.

As can be seen, the procedure proposed in Section 11.5 is sensitive enough to detect the disagreement between the Princeton experimental results and the theoretical predictions for velocity profiles in smooth pipes, which starts at the point where the roughness protrudes beyond the viscous sublayer, according to the drag coefficient data.

Moreover, consider the kinematic viscosity of air in the last run that corresponds to a smooth pipe. According to the Princeton data it can be estimated at approximately 1.05×10^{-2} cm^2/s, which is equal to the kinematic viscosity of water. (Remember that the Nikuradze (1932) experiments were performed on water flows in pipes.)

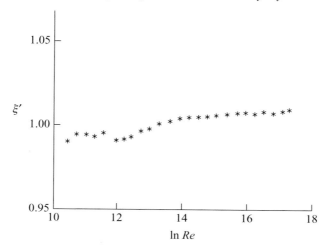

Figure 11.19 The relative friction coefficient $\xi' = \lambda_{\mathrm{exp}}/\lambda_{\mathrm{corrected}}$, with viscosity corrected as in Figure 11.16, does not exhibit a kink at $Re \sim 10^6$. From Barenblatt, Chorin and Prostokishin (1997).

We come to the conclusion that the Princeton group did not surpass the range of Reynolds number achieved by Nikuradze or reach its upper bound as far as accurate velocity measurements are concerned. However, these experiments of Dr Zagarola were of substantial importance for the following reasons. In contrast with the published Nikuradze data,[2] Zagarola's data contain measurements far from the envelope. This allowed a clear demonstration of the split in the data according to the Reynolds number. Also, Zagarola's data clearly revealed the linear part of the curves in the $(\ln \eta, \phi)$-plane corresponding to each Reynolds number. These effects were so robust that even an inaccuracy in the velocity measurements did not prevent them from being clearly revealed.

11.10 Scaling laws for turbulent boundary layers

By the logic of its derivation, the scaling law (11.30), (11.31) must be valid not only for flows in pipes but also for any wall-bounded shear flows.

Here, however, a basic question appears: what is the definition of the Reynolds number for these flows which would allow the use of the law (11.30), (11.31) for them? This basic question is immaterial as long as the engineer or researcher continues to believe in the universal logarithmic law. Indeed, if the law is *Re*-independent then the definition of *Re* does not matter. The situation is different,

[2] Zagarola's results call into question the completeness of Nikuradze's published data.

though, when the law is *Re*-dependent. But what should one do for wall-bounded shear flows other than pipe flow?

Below, we will consider boundary layers and will show that the law (11.30), (11.31) describes these flows also under an appropriate choice of Reynolds numbers. Zero-pressure-gradient boundary layers have been much investigated experimentally over the last 25 years. The common choice of Reynolds number for these flows is

$$ Re_\theta = \frac{U\theta}{\nu}, \qquad \theta = \frac{1}{U^2} \int_0^\infty u(U - u)\, dy, \qquad (11.68) $$

where U is the free-stream velocity and θ is a length calculated from integration of the velocity profile, the so-called *momentum thickness*. This choice is rather arbitrary, and a priori the law (11.30), (11.31) with $Re = Re_\theta$ should not be valid. Indeed, what is the proper choice of Re for the boundary layers?

To understand this, we have to confirm first that in the intermediate region of the boundary layer flow adjacent to the viscous sublayer *some scaling law is valid*. To do this, all the available experimental data presented in the traditional $(\ln \eta, \phi)$-plane were replotted in a bilogarithmic $(\ln \eta, \ln \phi)$-plane. The result was instructive: without any exception, for all investigated flows a straight line was obtained for region I, the region adjacent to the viscous sublayer (see the examples in Figure 11.20). Moreover, for flows with low free-stream turbulence a second self-similar region, II, was observed between the first region and the free-stream flow.

The basic question is whether a unique length scale Λ exists playing the same role for the intermediate region I of the boundary layer as does the diameter for pipe flow. In other words, is it possible to find a length scale Λ, perhaps influenced by individual features of the flow, such that the scaling law (11.30), (11.31) is valid for the first intermediate region I?

To answer this question two coefficients, A and α (obtained, we emphasize, by statistical analysis of the experimental data in the first intermediate scaling region, region I), were taken and two values, $\ln Re_1$ and $\ln Re_2$, were then calculated by solving two equations suggested by the scaling law (11.30):

$$ \frac{1}{\sqrt{3}} \ln Re_1 + \frac{5}{2} = A, \qquad \frac{3}{2 \ln Re_2} = \alpha. \qquad (11.69) $$

If the values of $\ln Re_1$ and $\ln Re_2$ obtained by solving these two different equations are indeed close, i.e. if they coincide to within experimental accuracy, it means that a unique length scale Λ can be determined and the experimental scaling law in region I coincides with the basic scaling law (11.30), (11.31).

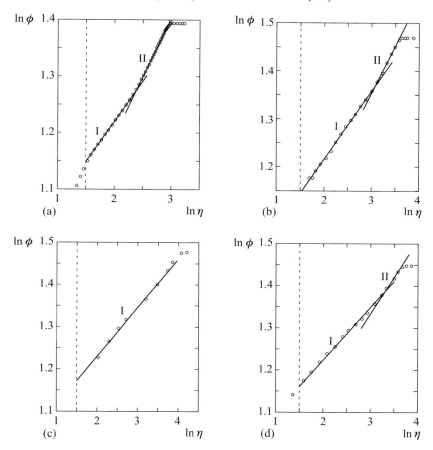

Figure 11.20 The data, replotted in bilogarithmic form, obtained from experiments in turbulent boundary layers. (a) The data of Erm and Joubert (1991); $Re_\theta = 2788$. Both self-similar intermediate regions, I and II, are clearly seen. (b) The data of Krogstad and Antonia (1999); $Re_\theta = 12\,570$. Again, both the self-similar intermediate regions, I and II, are clearly seen. (c) The data of Petrie, Fontaine, Sommer and Brugart obtained by scanning the graphs in Fernholz and Finley (1996); $Re_\theta = 35\,530$. The first self-similar region, I, is seen but the second self-similar region, II, is barely revealed. (d) The data of Smith obtained by scanning the graphs in Fernholz and Finley (1996); $Re_\theta = 12\,990$. The first self-similar region, I, and the second region, II, are clearly seen. From Barenblatt, Chorin and Prostokishin (2000).

The comparison revealed that the values of $\ln Re_1$ and $\ln Re_2$ which enter the relations (11.69) are indeed close. Thus, a mean Reynolds number, Re, can be introduced, for instance by writing

$$Re = \sqrt{Re_1 Re_2}, \qquad \ln Re = \tfrac{1}{2}(\ln Re_1 + \ln Re_2), \qquad (11.70)$$

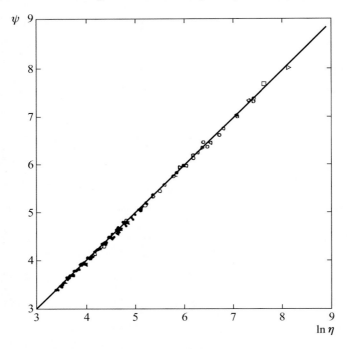

Figure 11.21 The data are as follows: ∗, Erm and Joubert (1991); Δ, Krogstad and Antonia (1999); □, Smith; and ∇, Petrie *et al*. All the data collapse onto the bisectrix of the first quadrant in the $(\ln \eta, \psi)$-plane, in accordance with the universal form (11.71) of the scaling law (11.31). From Barenblatt, Chorin and Prostokishin (2000).

so that the value of *Re* obtained in this way can be considered as an estimate for the effective Reynolds number of the boundary layer flow.

Checking the universal form of the scaling law (11.31),

$$\psi = \frac{1}{\alpha} \ln \left(\frac{2\alpha\phi}{\sqrt{3} + 5\alpha} \right) = \ln \eta, \qquad \alpha = \frac{3}{2 \ln Re}, \qquad (11.71)$$

where *Re* is obtained by means of the relation (11.70), gives another way of demonstrating clearly its applicability to the first intermediate region of the flow adjacent to the viscous sublayer. According to the relation (11.71) in the coordinates $\ln \eta, \psi$, all the experimental points should collapse onto the bisectrix of the first quadrant, as for the case of the pipe flow in Figure 11.7. The data seen in Figure 11.21 do indeed collapse onto the bisectrix with sufficient accuracy to confirm the scaling law (11.30), (11.31).

We conclude that the scaling law (11.30), (11.31) gives an accurate description of the mean velocity distribution over the self-similar intermediate region I adjacent to the viscous sublayer for a wide variety of zero-pressure-gradient boundary layer flows.

In a paper by Professor Panton (Panton, 2002) an important comment was made: "... the method that Barenblatt, Chorin and Prostokishin proposed to extend the power law to boundary layers displays extreme sensitivity". Following this comment, values of $\ln Re$ in the interval between the close values $\ln Re_1$ and $\ln Re_2$ that were different from the rather arbitrarily taken value (11.70) were tried and, indeed, a slightly better agreement with the scaling law was achieved. (The procedure for selecting the best value of $\ln Re$ lying between $\ln Re_1$ and $\ln Re_2$ for a given set of data is easily performed on a computer.) The selection of $\ln Re$ according to (11.70) was a first trial, and it allowed us to obtain satisfactory results for the available set of data.

We recall that the Reynolds number is defined as $Re = U\Lambda/\nu$, where U is the free-stream velocity and Λ is a length scale which is well defined for all the flows under investigation. An analysis of the experimental data showed that Λ is between 1.5 and 1.6 times the wall-region thickness as determined by the sharp intersection of the two velocity distribution laws I and II. The validity of the scaling law (11.30), (11.31) for the lower self-similar region, I, of boundary-layer flows constitutes a strong argument in favor of its validity for a wide class of wall-bounded turbulent shear flows at large Reynolds numbers.

The nature of the second self-similar region, II, adjacent to the basic stream, is not yet completely clear. For zero-pressure-gradient boundary layer flows in the absence of free-stream turbulence, the power β in the scaling law valid for this region,

$$\phi = B\eta^\beta , \tag{11.72}$$

is close to 0.2. The data for non-zero-pressure-gradient boundary layers are substantially less numerous. The processing of the data of Marušić and Perry (1995) confirmed the chevron-like structure of the velocity distribution of the boundary layer. It showed that the power β has a substantial variation (see Figure 11.22).

Let us determine the set of parameters that govern the coefficient B and the power β in the scaling law (11.72). One parameter must be the effective Reynolds number Re which determines the flow structure in layer I and is affected, in turn, by the flow in the viscous sublayer and in layer II. The following dimensional parameters should also influence the flow in the upper layer: the pressure gradient $\partial_x p$, the dynamic (friction) velocity u_* and the fluid properties, its kinematic viscosity ν and density ρ. The dimensions of the governing parameters in the LMT class are as follows:

$$[\partial_x p] = \frac{M}{L^2 T^2} , \qquad [u_*] = \frac{L}{T} , \qquad [\nu] = \frac{L^2}{T} , \qquad [\rho] = \frac{M}{L^3} .$$

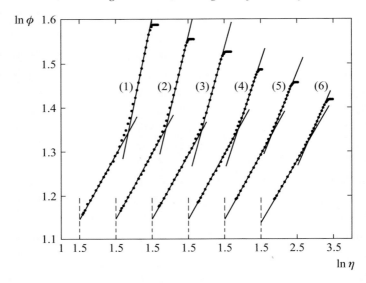

Figure 11.22 The mean velocity profiles in bilogarithmic coordinates in the series of experiments of Marušić and Perry (1995) for $U = 30$ m/s and an adverse pressure gradient. The "chevron" structure of the profiles is apparent and regions I and II are clearly distinguishable. From Barenblatt, Chorin and Prostokishin (2002).

(1)	$Re = 19\,133,$	$\ln Re_\Lambda = 8.83,$	$P = 7.04 \times 10^{-3},$	$\beta = 0.388;$
(2)	$Re = 16\,584,$	$\ln Re_\Lambda = 10.18,$	$P = 5.79 \times 10^{-3},$	$\beta = 0.346;$
(3)	$Re = 14\,208,$	$\ln Re_\Lambda = 10.20,$	$P = 4.2 \times 10^{-3},$	$\beta = 0.306;$
(4)	$Re = 10\,997,$	$\ln Re_\Lambda = 10.31,$	$P = 2.86 \times 10^{-3},$	$\beta = 0.247;$
(5)	$Re = 8\,588,$	$\ln Re_\Lambda = 10.323,$	$P = 1.75 \times 10^{-3},$	$\beta = 0.207;$
(6)	$Re = 6\,430,$	$\ln Re_\Lambda = 10.51,$	$P = 0,$	$\beta = 0.190.$

The last three parameters have independent dimensions. Therefore only one additional dimensionless governing parameter can be formed from the dimensional parameters:

$$P = \frac{\nu \partial_x p}{\rho u_*^3} .$$ (11.73)

We come to the conclusion that the power β and the coefficient B depend upon two dimensionless governing parameters, Re and P.

Summing up, we have shown that the von Kármán–Prandtl universal logarithmic law for the intermediate region of wall-bounded turbulent shear flow is not quite correct. It must be abandoned and replaced by a power law for the velocity distribution (11.30), (11.31) and that which directly follows from it, without any ad hoc manipulation with constants, the friction law (11.38).

12

Turbulence: mathematical models of turbulent shear flows and of the local structure of turbulent flows at very large Reynolds numbers

12.1 Basic equations for wall-bounded turbulent shear flows. Wall region

Consider the Reynolds equation (11.13) (which is, we remind the reader, the averaged Navier–Stokes equation):

$$\partial_t \rho \bar{\mathbf{u}} + \rho (\bar{\mathbf{u}} \cdot \nabla) \cdot \bar{\mathbf{u}} = -\nabla \bar{p} + \nabla \cdot (\bar{\boldsymbol{\tau}} + \mathbf{T}) \,,$$

where $\mathbf{u} = (u, v, w)$. It is easy to show that for a shear flow this vector equation is reduced to three scalar equations:

$$-\partial_x \bar{p} + \frac{d}{dz} (\bar{\tau}_{xz} + T_{xz}) = 0 \,, \tag{12.1}$$

$$\partial_y \bar{p} = \partial_z \bar{p} = 0 \,. \tag{12.2}$$

For a shear flow we have $\bar{\tau}_{xz} = \eta (du/dz)$, $T_{xz} = -\rho \overline{u'w'}$. Also, owing to the relation (12.2) $\partial_x \bar{p}$ is constant: according to (12.1) it can depend only upon z but $\partial_z \bar{p}$ is equal to zero. By integrating (12.1) we obtain for the total shear stress

$$\tau_{xz} + T_{xz} = \eta \frac{du}{dz} - \rho \overline{u'w'} = (\partial_x \bar{p}) z + \text{const} \,. \tag{12.3}$$

Put $z = 0$ in (12.3) to obtain

$$\text{const} = \left(\eta \frac{du}{dz} - \rho \overline{u'w'} \right)_{z=0} = \tau \,, \tag{12.4}$$

where τ is the shear stress at the wall. In the previous chapter the dynamic or friction velocity u_* was introduced, so in terms of this the constant equals ρu_*^2.

The viscous component of the shear stress $\eta (du/dz)$ is comparable with the Reynolds stress $-\rho \overline{u'w'}$ created by turbulent vortices only in the close vicinity – the viscous sublayer – of the wall $z = 0$, which is the flow boundary. The flow in the viscous sublayer will not be considered: up to now it has not been completely clarified. However, as we will see further, the remaining part of the shear flow can be considered independently, and for computing the discharge (the fluid mass flow through the pipe cross-section) and the drag it is the basic part. At the same time only the intermediate region of the shear flow, where z is sufficiently small, will be considered, so that the term $(d\bar{p}/dx)z$ can be neglected also.

So, we consider below the intermediate *wall region* of a wall-bounded turbulent shear flow, where the viscous stress and the direct contribution of the pressure gradient can be neglected. This intermediate region is important, and the flow structure in it can be satisfactorily clarified experimentally and then modeled.

In this wall region of the turbulent shear flow the momentum balance equation takes the form

$$-\rho\overline{u'w'} = \rho u_*^2 . \tag{12.5}$$

Of fundamental importance for the future analysis of the wall region of the turbulent shear flow is the balance equation for the energy of turbulence, i.e. the energy of the turbulent vortices. To derive this equation it is necessary to perform the following steps in order:[1]

(1) multiply the Navier–Stokes equation by the velocity and average the equation obtained;
(2) derive the equation of energy for the *mean flow* by multiplying the Reynolds equation by the average velocity;
(3) subtract the equation thus obtained from that obtained in step 1. Then, using the simplifications discussed above for shear flow, we obtain the balance equation for the turbulence energy:

$$\overline{(-\rho u'w')}\frac{du}{dz} - \rho\varepsilon = 0 . \tag{12.6}$$

The first term on the left-hand side of (12.6) represents the rate of inflow of turbulence energy – the energy of the turbulent vortices – from the mean flow per unit volume. With the opposite sign this term enters the energy balance equation for the mean flow obtained in step 2: this clarifies its physical sense. Furthermore, $\rho\varepsilon$ is the rate of turbulence energy dissipation into heat per unit volume:

$$\rho\varepsilon = \frac{\eta}{2}\overline{(\partial_\alpha u'_\beta + \partial_\beta u'_\alpha)(\partial_\alpha u'_\beta + \partial_\beta u'_\alpha)} . \tag{12.7}$$

(Here summation over repeated Greek indices from 1 to 3 is assumed and also we have set $\bar{u}_1 = u$, $u'_1 = u'$, $u'_2 = v'$, $u'_3 = w'$, $x_1 = x$, $x_2 = y$, $x_3 = z$.)

In equation (12.6) we have neglected the term

$$\rho\frac{d}{dz}\left(-\frac{1}{2}\overline{u'_\beta u'_\beta u'_3} + \overline{\nu u'_\beta(\partial_\beta u'_3 + \partial_3 u'_\beta)} - \frac{1}{\rho}\overline{p'u'_3}\right) , \tag{12.8}$$

---

[1] We recommend that readers perform all these simple steps. It will give them a "hands-on" feeling for the problem.

representing the viscous and turbulent transfer of the turbulence energy and the work of the pressure fluctuations on the velocity fluctuations. This neglect seems to be possible in the wall region of the turbulent shear flow but not in the viscous sublayer. Discussion of this term and of the possibility of neglecting it in the wall region can be found in more detail in the book by Monin and Yaglom (1971), pp. 381–8. Kolmogorov (1942) and Prandtl (1945) proposed using the equation for turbulence energy balance (12.6) to complete the Reynolds equation (12.5), in their endeavours to obtain a closed system of relations for turbulent flows. This proposal had far-reaching consequences.

12.2 Kolmogorov–Prandtl semi-empirical model for the wall region of a shear flow

Now we introduce the coefficient of turbulent momentum exchange, the kinematic eddy viscosity

$$k = \frac{-\rho \overline{u'w'}}{\rho \, du/dz} \, . \tag{12.9}$$

We emphasize that the introduction of this scalar eddy viscosity k *for a shear flow* is not a new hypothesis; we have merely defined a new quantity. Equations (12.5), (12.6) take the form:

$$k \frac{du}{dz} = u_*^2 \, , \qquad k \left(\frac{du}{dz} \right)^2 - \varepsilon = 0 \, . \tag{12.10}$$

Here, we recall that $u_* = (\tau/\rho)^{1/2}$ is the governing parameter of shear flow: it is the dynamic or friction velocity.

The basic hypothesis underlying the shear flow model of Kolmogorov (1942) and of Prandtl (1945) can be formulated in the following way. *At large Reynolds numbers, the local structure of the field of vortices around any point far from the boundaries is statistically identical for all shear flows at a given Reynolds number, so that only the scales of time and space are different for different flows.* Therefore, for a given Reynolds number all dimensionless flow properties should be identical. This means that all kinematic flow properties at a point, including the kinematic eddy viscosity k and the mean dissipation rate per unit mass ε, are determined only by the local values of any two kinematic properties having different dimensions. The mean turbulence energy per unit mass,

$$b = \frac{\overline{u'^2 + u'^2 + w'^2}}{2} \, ,$$

and the external length scale ℓ (the mean length scale of the vortices) can be selected as such properties; this gives the ℓ, b version of the Kolmogorov–Prandtl model. Also in wide use is a b, ε version, where the quantities b and ε, the energy dissipation rate per unit mass, are selected as basic. In principle the two versions are logically equivalent. In what follows we will use Kolmogorov's b, ℓ version, preferred by Kolmogorov himself and his school (see Monin and Yaglom, 1971).

Dimensional analysis easily gives the following relations:

$$k = \ell \sqrt{b}, \qquad \varepsilon = \gamma^4 \frac{b^{3/2}}{\ell}. \tag{12.11}$$

The coefficient in the first equation of (12.11) can be set equal to 1 because the length scale of vortices is determined only up to a constant factor. The constant γ is, in principle, a Reynolds-number-dependent quantity. At large Reynolds numbers this quantity is close to 0.5 (see Monin and Yaglom, 1971).

Thus, the balance equations for momentum and turbulence energy (12.10) take the form

$$\ell \sqrt{b} \frac{du}{dz} = u_*^2, \qquad \ell \sqrt{b} \left(\frac{du}{dz}\right)^2 - \gamma^4 \frac{b^{3/2}}{\ell} = 0. \tag{12.12}$$

It is instructive that from equations (12.12), without any further assumptions concerning the length scale ℓ, the relation for the turbulence energy in the intermediate wall region of a wall-bounded turbulent shear flow follows:

$$b = \frac{u_*^2}{\gamma^2}. \tag{12.13}$$

Relation (12.13) is important: in particular it shows that the dynamic velocity u_* determines the magnitude of the velocity fluctuations in the wall region. Indeed, assuming the isotropy of the velocity fluctuations in the wall region as an approximation we obtain

$$\overline{u'^2} \sim \frac{2}{3\gamma^2} u_*^2, \qquad \sqrt{\overline{u'^2}} \sim \frac{u_*}{\gamma}. \tag{12.14}$$

Obviously the system (12.12) is not closed, because the length scale ℓ is unknown. However, when ℓ is determined in some way, system (12.12) becomes a basic tool used in many practical calculations, especially in the physics of the atmosphere and the ocean (Monin and Yaglom, 1971). In particular, if ℓ is determined, the first equation of (12.12) determines the gradient of the velocity distribution.

The situation regarding the determination of the length scale ℓ is, however, non-trivial. Indeed, a relation for ℓ can be obtained using dimensional analysis, assuming that ℓ depends on the distance from the wall z (the thickness of the viscous sublayer is considered negligible), the dynamic velocity u_*, the kinematic viscosity ν and the Reynolds number: $\ell = f(z, u_*, \nu, Re)$. The standard procedure of dimensional analysis gives

$$\ell = z \, \Phi \left(Re, \, \frac{u_* z}{\nu} \right) . \tag{12.15}$$

We now remember that the intermediate wall region of the shear flow at large Reynolds numbers is being considered, where $u_* z/\nu$ is also large. Therefore, the classical approach, like the direct application of the similarity approach used in the determination of the velocity gradient (see the previous chapter) is based on the assumption that the function Φ can be replaced by its limit $\kappa\gamma = \Phi(\infty, \infty)$, which is finite and Reynolds-number independent (see e.g. Monin and Yaglom, 1971). In this way the von Kármán–Prandtl universal logarithmic law appears:

$$\frac{du}{dz} = \frac{u_*^2 \gamma}{\kappa \gamma u_* z} = \frac{u_*}{\kappa z} , \qquad \frac{u}{u_*} = \frac{1}{\kappa} \ln z + C' .$$

However, with the support of much experimental data we proved unambiguously in the previous chapter that this is not quite correct. In fact, at large Reynolds numbers the velocity distribution is represented not by a single universal curve but, as was shown, by a family of Reynolds-number-dependent scaling laws:

$$\frac{u}{u_*} = \left(\frac{1}{\sqrt{3}} \ln Re + \frac{5}{2} \right) \left(\frac{u_* z}{\nu} \right)^{3/(2 \ln Re)} . \tag{12.16}$$

Moreover, according to the relation (11.51) in the middle range, II, of the parameter $\eta = u_* z/\nu$ (see Figure 11.12 and also Zagarola's graph in Figure 11.6), the family of velocity distributions (12.16) can be represented by a family of Reynolds-number-dependent non-universal logarithmic laws:

$$\frac{u}{u_*} = \frac{1}{\kappa(Re)} \ln \frac{u_* z}{\nu} + B(Re) , \tag{12.17}$$

where

$$\kappa(Re) = e^{-3/2} \left(\sqrt{\frac{3}{2}} + \frac{15}{4 \ln Re} \right)^{-1} , \tag{12.18}$$

$$B(Re) = -\frac{e^{3/2} \ln Re}{2\sqrt{3}} - \frac{5}{4} e^{3/2} . \tag{12.19}$$

Using (12.13) and (12.16) we obtain, for the length scale ℓ from the first equation (12.12),

$$\ell = \frac{\gamma z}{\left[\sqrt{3}/2 + 15/(4 \ln Re) \right]} \left(\frac{u_* z}{\nu} \right)^{-3/(2 \ln Re)} . \tag{12.20}$$

Thus, we have demonstrated that ℓ is proportional not to z but to $z^{1-3/(2 \ln Re)}$, so that there is incomplete similarity in the parameter $\eta = u_* z/\nu$ in the relation (12.15) at large Re and η. We repeat that this conclusion is strongly supported by experiment. It has an important consequence. It shows that at finite viscosity ν the transverse vortical structures in the turbulent shear flow are not space filling. Complete similarity in (12.15) would mean, as is emphasized in traditional arguments, that the vortical structures are space filling and that the length scale is proportional to z.

In a viscous flow one can define essential support of the vorticity (Chorin, 1994) in the regions where the absolute vorticity exceeds a certain predetermined threshold. We are now able to appreciate properly the prophetic words of Prandtl:

> For lower Reynolds numbers the agreement [with the von Kármán formulae] is worse, and this can be attributed to the action of the viscosity also in the inner part of the flow, i.e. to the viscosity-influenced streaks of which the laminar layer at the wall consists [in fact, the flow in this layer is also turbulent; therefore the term "viscous sublayer" is used nowadays] and which in this case enter far into the internal part of the flow.

These "streaks" are exactly the regions where the absolute value of the vorticity is high. In the late 1960s, in experiments of S. J. Kline, W. J. Reynolds and their followers (Kline *et al.*, 1967) and those of later workers, these streaks were observed and it was shown that they have a width of order 100 ν/u_*.

12.3 A model for drag reduction by polymeric additives

Drag reduction and the mean velocity increase in a turbulent shear flow through the addition of tiny amounts (several parts per million) of high-molecular-weight polymers (the Thoms effect) has been known for a long time, in fact since World War II. More recently, it has been understood that this effect is related to supramolecular structures formed in the flow, consisting of polymer molecules with solvent molecules attached to them. Recent experiments by S. Chu, E. S. G. Shaqfeh, R. E. Teixeira and their associates, where the motion of supramolecular structures was directly observed, have made it possible to understand and quantify the dynamic interaction of the polymeric structure and the solvent (water) flow. These results have led to the construction of a mathematical model of the Thoms effect based on the Kolmogorov–Prandtl semi-empirical model of shear flow turbulence presented in Section 12.2 (see Barenblatt, 2008, and references therein), which is presented below.

It is most important to note that water with polymeric additives having the property of turbulent drag reduction does not constitute a genuine solution: supramolecular polymeric structures attaching the solvent molecules are formed in the mixture. The role of such structures in the Thoms effect was suggested rather a long time ago. An indirect confirmation of the viscoelastic behaviour of polymeric supramolecular structures was obtained in the paper Kudin *et al.* (1973). A strong water jet directed onto a metallic plate did not affect it for several hours. However, the addition of tiny amounts of polymer led to a fast abrasive-like wearing of the plate.

However, only in the work of Chu, Shaqfeh, Teixeira and their associates (Teixeira *et al.*, 2005, 2007) complicated "tumbling" motions of the supramolecular structures were observed. These motions were accompanied by the deformation, overturning, rotation and adjustment of the elongated structures to the local flow. The authors introduced and measured the characteristic time of tumbling, the "disentanglement time" θ.

These experimental studies suggest the basic hypothesis of the proposed model: *The force* **F** *acting on the solvent (water) from the supramolecular polymeric structures, which consist of networks of polymeric molecules with attached solvent molecules, is proportional to the concentration s of the supramolecular structures. It is also determined by the instantaneous velocity fluctuation,* **u′** = **u** − **ū**, *and the disentanglement time* θ.

Dimensional analysis and symmetry considerations (among the governing parameters only **u′** is the vector; therefore the force **f** should be directed along **u′**) give

$$\mathbf{F} = -As\,\frac{\mathbf{u}'}{\theta}\,. \tag{12.21}$$

The minus sign is due to the obvious fact that the force is a reaction to the velocity fluctuation. Also, A is a dimensionless constant which in principle could be included in the time θ. That has not been done here, bearing in mind that the disentanglement time was independently introduced and measured by the experimentalists.

To summarize: for a dilute solution, when the interaction of the supramolecular structures can be neglected and the density of the solution can be assumed to be equal to the density of the solvent, the momentum balance equation in (12.12) for the solvent flow in the presence of supramolecular structures remains the same as for pure solvent flow.

The situation is different as far as the equation of turbulent energy balance for the flow of solvent is considered. We emphasize that the flow of the solvent, not that of the whole mixture, is considered.

Repeating the derivation of the turbulence energy balance (12.6), now including the mass force acting on the solvent, we obtain

$$(-\rho\overline{u'w'})\,\frac{du}{dz} - \rho\varepsilon - \rho\overline{\mathbf{u}' \cdot \mathbf{F}'} = 0\,. \tag{12.22}$$

Taking into account (12.9), (12.11) and (12.21) we can reduce equation (12.22) to the form

$$\ell\,\sqrt{b}\left(\frac{du}{dz}\right)^2 - \gamma^4\,\frac{b^{3/2}}{\ell} - \frac{2Asb}{\theta} = 0\,, \tag{12.23}$$

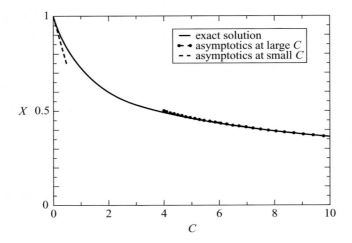

Figure 12.1 Dependence of X on the parameter C; see the text.

because obviously $\overline{u'^2} = \overline{u'_\alpha u'_\alpha} = 2b$. The third term represents the work done by the turbulent vortices in disentangling the supramolecular structures. Denoting $\gamma b^{1/2}/u_*$ by X, we obtain for X an algebraic equation of the fourth degree:

$$X^4 + 2CX^3 - 1 = 0, \tag{12.24}$$

where the dimensionless parameter C is given by

$$C = \frac{As}{\gamma^3} \frac{1}{u_*\theta}. \tag{12.25}$$

The first equation in (12.12), which remains valid for the flow of the solvent, and equation (12.17) show that in the middle range of the parameter $\eta = u_*z/\nu$ the following relation holds:

$$\ell = \kappa(Re)\gamma z. \tag{12.26}$$

We now remember that $\kappa(Re)$ is much less than the value $\kappa \sim 0.4$ usually assumed for the von Kármán constant. Therefore, owing to (12.25),

$$C = \frac{\kappa(Re)Asz}{\gamma^2 u_*\theta}. \tag{12.27}$$

The plot of $X = \gamma b^{1/2}/u_*$ as a function of C is presented in Figure 12.1; it illustrates the decrease in turbulent energy. For the momentum exchange coefficient $k = \ell \sqrt{b}$ we obtain $k = \kappa(Re)zu_\kappa X(C)$, and the mean velocity is obtained by integration,

$$u = \frac{u_*}{\kappa(Re)} \int \frac{dz}{zX(C)} = \frac{u_*}{\kappa} \int \frac{dC}{CX(C)}. \tag{12.28}$$

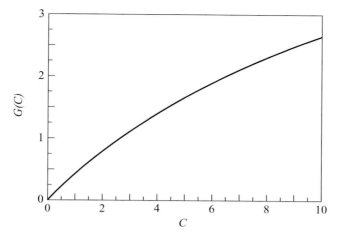

Figure 12.2 The integral $G(C)$ as a function of C; see the text.

In the case $C \ll 1$ (corresponding to a small polymer concentration or a large disentanglement time θ), equation (12.24) gives $X(C) = 1 - C/2$, so, according to (12.28),

$$u = \frac{u_*}{\kappa(Re)} \ln \frac{z}{u_*\theta} + \frac{Asz}{2\theta\gamma^2} + \text{const} \qquad (12.29)$$

or

$$u = \frac{u_*}{\kappa(Re)} \ln \frac{u_*z}{\nu} + \frac{Asz}{2\theta\gamma^2} + \text{const}_1 . \qquad (12.30)$$

Thus we can see that the polymeric additives lead to an additional linear term in the expression for the velocity.

In the case $C \gg 1$ the first term in equation (12.24) is small in comparison with the second, and we get $X \sim (2C)^{-1/3}$. This leads to a power-law velocity profile,

$$u \sim 3 \left[2\kappa(Re)As/\gamma^2 \right]^{1/3} u_*^{2/3} \theta^{-1/3} z^{1/3} + \text{const} .$$

For intermediate values of the parameter C the velocity distribution is represented in the form

$$u = \frac{u_*}{\kappa(Re)} \ln \frac{u_*z}{\nu} + \frac{u_*}{\kappa(Re)} G \left(\frac{\kappa(Re)Asz}{\gamma^2 u_*\theta} \right) + \text{const} ,$$

where

$$G(C) = \int_0^C \frac{1 - X(C)}{CX(C)} \, dC .$$

A plot of the function $G(C)$ is given in Figure 12.2. The constants in all these formulae depend on the Reynolds number and as Re tends to ∞ they tend to $-\infty$.

12.4 The local structure of turbulent flows at very large Reynolds numbers

The mathematical model of the local structure of turbulent flows at very high Reynolds numbers is generally considered to be the supreme achievement of theoretical turbulence studies in the twentieth century. It was created originally by A. N. Kolmogorov (1941a, b) and A. M. Obukhov (1941a, b), at that time Kolmogorov's graduate student. Independently, but later, the eminent physicists L. Onsager (1945), W. Heisenberg (1948) and C. von Weizsäcker (1948) arrived at the same idea. It was a remarkable example of congeniality in the advancing of a fundamental problem; similarly, we can remember and compare the practically simultaneous discovery of non-Euclidean geometry by Gauss, Lobachevsky and Bolyai.

In this section a concise presentation of this model will be given, omitting some technical details, for which references will be given. Basically we will follow the original ideas of Kolmogorov and Obukhov. In general, however, our approach will be somewhat different. It will be based on a deep analogy between the local structure of general turbulent flows at very large Reynolds numbers and the wall-region structure of developed turbulent shear flows. Detailed, well-established, experimental results for shear flows will be used for comparison. The analogy between local structure and shear flow was understood and emphasized by Chorin (1977).

So, arbitrary turbulent flows at very large Reynolds numbers will be considered. Such flows are difficult, or even impossible, to obtain in the laboratory; we can expect to find important applications of a model of local structure in terrestrial, atmospheric and/or oceanic flows. In the early 1950s the Soviet astronomer G. Shain made an attempt to test the Kolmogorov–Obukhov theory for stellar constellations. Nevertheless, in our presentation we preserve the approximation of incompressible constant-density Newtonian fluid.

The basic assumption of the model is the following hypothesis. *All turbulent flows at very large Reynolds numbers have a universal, although Reynolds-number-dependent, local structure.* As in the case of turbulent shear flow, presented in the previous section, this means that, statistically, the local geometric and kinematic features of turbulent vortices are identical for all turbulent flows at a given Reynolds number; only the time and length scales are different.

From this basic assumption it follows that, locally, at very large Reynolds numbers all turbulent flows are homogeneous and isotropic: indeed, among turbulent flows are fully isotropic and homogeneous flows which possess exactly the property of *complete* isotropy and homogeneity and, consequently, the property of *local* isotropy and homogeneity. It was Kolmogorov (1941a, b) who introduced implicitly, and used, the hypothesis of universality.

Consider now a turbulent flow field in a region Ω in the vicinity of an arbitrary point **x**. We denote by $\partial\Omega$ the boundary of this region. The linear scale of Ω is assumed to be small in comparison with the external scale of the flow Λ. At the same time the time scales of the fluctuations in Ω are assumed to be small in comparison with the time scale of the basic external flow, so that the flow in Ω can be considered to be statistically steady.

Owing to the local homogeneity the mean velocity in Ω is constant:

$$\bar{\mathbf{u}}(\mathbf{x} + \mathbf{r}) = \bar{\mathbf{u}}(\mathbf{x}), \tag{12.31}$$

where $\mathbf{r}$ is the radius vector of an arbitrary point in Ω in a system with its origin at $\mathbf{x}$. Owing to (12.31) the first-order correlation moment $[\overline{\mathbf{u}(\mathbf{x} + \mathbf{r}) - \mathbf{u}(\mathbf{x})}]$ is equal to zero. The second-order correlation moment is given by the second-rank tensor

$$\mathbf{D} = \overline{[\mathbf{u}(\mathbf{x} + \mathbf{r}) - \mathbf{u}(\mathbf{x})] \otimes [\mathbf{u}(\mathbf{x} + \mathbf{r}) - \mathbf{u}(\mathbf{x})]}, \tag{12.32}$$

whose components are

$$D_{ij} = \overline{[u_i(\mathbf{x} + \mathbf{r}) - u_i(\mathbf{x})][u_j(\mathbf{x} + \mathbf{r}) - u_j(\mathbf{x})]} . \tag{12.33}$$

An easy but lengthy calculation, which can be found in e.g. Landau and Lifshitz (1987, pp. 135–7), shows that all the components of the tensor $\mathbf{D}$ can be expressed for the locally isotropic and homogeneous flow of an incompressible fluid as functions of just one, of them, say,

$$D_{11} = \overline{[u_1(\mathbf{x} + \mathbf{r}) - u_1(\mathbf{x})]^2} ; \tag{12.34}$$

thus

$$D_{ij} = D_{11}\delta_{ij} + \left(\frac{r}{2} \frac{dD_{11}}{dr}\right)(\delta_{ij} - \xi_i\xi_j) . \tag{12.35}$$

Here $\xi_i = r_i/r$, where $r = |\mathbf{r}|$ and r_i are the components of the vector $\mathbf{r}$ in the local orthonormal Cartesian system, such that $i = 1$ corresponds to the direction of $\mathbf{r}$ and $i = 2, 3$ are arbitrary mutually perpendicular directions that are perpendicular to $\mathbf{r}$. Owing to the local isotropy and incompressibility, it follows from (12.35) that

$$D_{22} = D_{33} = D_{11} + \frac{r}{2} \frac{dD_{11}}{dr} . \tag{12.36}$$

Similarly, for the components of the third-order correlation moment tensor we have

$$D_{ijk} = \overline{[u_i(\mathbf{x} + \mathbf{r}) - u_i(\mathbf{x})][u_j(\mathbf{x} + \mathbf{r}) - u_j(\mathbf{x})][u_k(\mathbf{x} + \mathbf{r}) - u_k(\mathbf{x})]} . \tag{12.37}$$

All components of this tensor can be expressed via the component

$$D_{111} = \overline{[u_1(\mathbf{x} + \mathbf{r}) - u_1(\mathbf{x})]^3} . \tag{12.38}$$

A simple but rather lengthy proof of this statement can be found for example on pp. 137–9 of Landau and Lifshitz (1987).

Furthermore, using the universality of the local structure and assuming full (not only local) isotropy and homogeneity of the flow, Kolmogorov (1941b) derived the relation between D_{111} and D_{11}:

$$D_{111} = -\frac{4}{5}\varepsilon r + 6\nu \frac{dD_{11}}{dr}. \tag{12.39}$$

In the derivation of this equation Kolmogorov (1941b) used the averaged Navier–Stokes equation obtained by von Kármán and Howarth (1938) for fully isotropic and homogeneous turbulent flow. Here, however, a challenge appeared. In the paper Kolmogorov (1941b) it was claimed that the assumption of complete isotropy and homogeneity is not necessary, so that equation (12.39) could be derived without this assumption. In his paper (Batchelor, 1947) that clarified and explained in detail the Kolmogorov–Ohukhov theory, Batchelor mentioned that this was the only argument in Kolmogorov's papers (1941a, b) that he was unable to reconstruct. Up to this day this challenge remains unresolved.

At the next stage the fundamental idea of a "vortex cascade", proposed by the British physicist L. F. Richardson (1922), becomes of crucial importance.[2]

According to this model, the system of vortices in a region Ω remains statistically steady. Large vortices enter the region Ω, crossing its boundary $\partial\Omega$, and start to generate smaller vortices. There are several plausible mechanisms whereby larger vortices generate smaller ones. Neither Richardson, nor Kolmogorov, who used substantially this idea, made these mechanisms more precise. However, it is significant that *according to Richardson's idea, larger vortices generate smaller vortices statistically of a length scale of the same order of magnitude.*

The smaller vortices generate even smaller vortices, etc. It is significant that, according to the vortex cascade idea, viscous dissipation plays a role only in the comparatively thin regions between vortices. These regions, where an elevated vorticity is concentrated, are analogous to the streaks that show up in the basic shear flow from the viscous sublayer, whose role was emphasized by Prandtl (see Section 11.4). Therefore, the energy of the larger vortices is transferred to smaller vortices of the next generation without noticeable losses.

Owing to the statistical steadiness of the flow in Ω, the flux of energy down the cascade of vortices is constant and is equal to the rate ε of energy dissipation into heat. However, this energy dissipation into heat is performed by small vortices. The scale of these vortices λ is determined by the fluid kinematic viscosity ν and by the dissipation rate per unit mass, which is equal to the energy flux down the cascade of the vortices. Here the assumption that the vortices generate smaller vortices having the same length scale becomes significant.

[2] Julian, Lord Hunt of Chesterton, the great-nephew of L. F. Richardson, is an eminent fluid dynamicist. He co-edited a special anniversary issue of the *Proceedings of the Royal Society of London* (Hunt *et al.*, 1991).

The length scale λ was determined by Kolmogorov and is named *Kolmogorov's length scale*. Dimensional analysis allowed Kolmogorov to determine λ to within an order of magnitude:

$$\lambda = \frac{\nu^{3/4}}{\varepsilon^{1/4}} \, . \tag{12.40}$$

The dimension of ε is obviously equal to L^2/T^3. The length scale λ has its analogue in shear flow as the viscosity length scale $\delta = \nu/u_*$.

Summing up, we are interested in finding the statistical quantity D_{11} in the region Ω as an intermediate asymptotics valid in the cascade of vortex scales covering the "inertia range"

$$\Lambda \gg r \gg \lambda \, , \tag{12.41}$$

where, we recall, Λ is the external length scale of the flow. The length scales of order Λ form the "energy range"; the length scales of order λ and less form the "dissipation range".

The general picture of the process in the inertia range is as follows. Larger vortices generate smaller ones, which have a length scale of the same order of magnitude. The smaller vortices transfer the energy to even smaller ones without substantial loss, and so it goes on until vortices of "dissipation scale" are reached, which dissipate the energy to heat. Between the vortices of each generation there are thin regions of elevated vorticity that are analogous to the streaks coming from the viscous sublayer in the shear flows.

This picture makes it natural to follow exactly the steps used earlier in the analysis of shear flow.

The correlation moment D_{11} depends on the following dimensional parameters: r, the modulus of the radius vector $\mathbf{r}$, the kinematic viscosity ν and the energy dissipation rate ε as well as a dimensionless parameter, the global Reynolds number of the flow, Re:

$$D_{11} = D_{11}(Re, r, \nu, \varepsilon) \, ; \tag{12.42}$$

we have $[\nu] = L^2 T^{-1}$, $[\varepsilon] = L^2 T^{-3}$, $[r] = L$. The standard procedure of dimensional analysis gives

$$D_{11} = (\varepsilon r)^{2/3} \, \Phi_{11}\left(Re, \frac{r}{\lambda}\right) \, . \tag{12.43}$$

We now introduce two basic hypotheses similar to those proposed for shear flow.

First hypothesis There is incomplete similarity in the parameter r/λ and no similarity in Re. Therefore D_{11} assumes the form

$$D_{11} = A(Re)(\varepsilon r)^{2/3}\left(\frac{r}{\lambda}\right)^{\alpha(Re)} \, . \tag{12.44}$$

Second hypothesis The vanishing-viscosity principle is assumed, according to which there exists a well-defined limit of D_{11} when the viscosity ν tends to zero.

Following exactly the same steps as in Section 11.5 for shear flow, we find that the functions $A(Re)$ and $\alpha(Re)$ can be represented as follows:

$$A(Re) = A_0 + A_1 \vartheta(Re) ; \qquad \alpha(Re) = \alpha_0 + \alpha_1 \vartheta(Re) . \qquad (12.45)$$

Here $\vartheta(Re)$ is again a small parameter, which tends to zero as $Re \to \infty$, and A_0, A_1, α_0 and α_1 are universal Reynolds-number-independent constants. Thus, the relation (12.43) can be represented in the following form:

$$D_{11} = (A_0 + A_1 \vartheta)(\varepsilon r)^{2/3} \exp[(\alpha_0 + \alpha_1 \vartheta) \ln(r/\lambda)] . \qquad (12.46)$$

According to the vanishing-viscosity principle α_0 should vanish, otherwise a well-defined limit of D_{11} as $\nu \to 0$, i.e. $Re \to \infty$, will not exist.

Still repeating the argument used in Section 11.5 for the shear flow, we obtain that for the local structure the expression for the small parameter ϑ is

$$\vartheta = \frac{1}{\ln Re} . \qquad (12.47)$$

The reason is that, on the one hand, if ϑ goes to zero faster than (12.47) then the expression (12.44) for D_{11} loses its dependence on the Reynolds number, which contradicts the experimental data. On the other hand, if ϑ goes to zero slower than (12.47) then the vanishing-viscosity principle will be violated; indeed, λ is defined by the relation (12.40) and in this case the factor $\ln \lambda$ would not be compensated (see also the important work of Castaing, Gagne and Hopfinger (1990), where the term $1/\ln Re$ appeared in calculations of the probability densities). We obtain finally (Barenblatt and Goldenfeld, 1995)

$$D_{11} = \left(A_0 + \frac{A_1}{\ln Re}\right)(\varepsilon r)^{2/3} \left(\frac{r}{\lambda}\right)^{\alpha_1 / \ln Re} . \qquad (12.48)$$

Kolmogorov's relation (12.39) gives

$$D_{111} = -\frac{4}{5} \varepsilon r + 6\nu \frac{dD_{11}}{dr}$$

$$= \varepsilon r \left[-\frac{4}{5} + 6\nu \left(A_0 + \frac{A_1}{\ln Re}\right) \varepsilon^{-1} r^{-4/3 + \alpha_1 / \ln Re} \right] \qquad (12.49)$$

$$= -\frac{4}{5} \varepsilon r + O\left[\left(\frac{\lambda}{r}\right)^{4/3 + \alpha_1 / \ln Re} \right] .$$

The second term on the right-hand side of (12.49) is negligible in the inertia range, so the final Kolmogorov Reynolds-number-independent (please note!) relation for D_{111} appears as

$$D_{111} = -\frac{4}{5} \varepsilon r . \qquad (12.50)$$

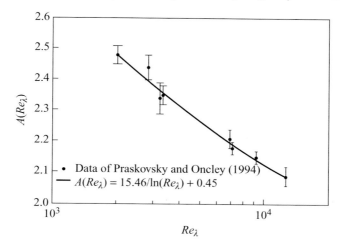

Figure 12.3 The Kolmogorov constant $A(Re)$ as measured by Praskovsky and Oncley (1994) plotted on a logarithmic scale against Taylor microscale Reynolds number Re_λ.

So, we have arrived at expressions (12.48) and (12.50) for the basic statistical characteristics of the local structure of turbulent flows at very large Reynolds numbers.

The experimental verification of these basic relations for the local structure of such a flow is very much poorer than the experimental verification of the basic relations for shear flows. In fact, experiments are more or less in agreement with the limiting value 2/3 of the exponent in the relation (12.48). This means that the correction $\alpha_1 / \ln Re$ is negligible, in contrast with the shear flow case. However, the experiments showed a significant Re-dependence of the pre-monomial constant; cf. $A(Re)$ in (12.44). We mention the correlation obtained by the processing of experimental measurements for the flow in a large wind tunnel by Praskovsky and Oncley (1994); see Figure 12.3.

This processing demonstrated that, for realistic values of the Reynolds numbers, the Reynolds-number-dependent term in the pre-power constant $A_1 / \ln Re$ is much larger than A_0; it was found that $A_0 = 0.45$ and $A_1 = 15.46$. This means that even oceanic, atmospheric or terrestrial experiments could be insufficient for obtaining reliable values of A_0 and A_1; the correlation data obtained by Praskovsky and Oncley (1994) can be used for a qualitative estimate only.

Let us make an important comment. The limiting form of the r-dependence for the second moment D_{11} is, according to (12.53), proportional to $r^{2/3}$. For the third moment (relation (12.50)) it is proportional to r. Some influential authors, for example, U. Frisch (see his book *Turbulence. The Legacy of A. N. Kolmogorov* (1996),

p. 132), have extrapolated these dependences to larger n as a law of $n/3$. Moreover, they have attributed this extrapolated law to A. N. Kolmogorov (1941a, b). Experimental data of various authors for a higher correlation moment have been reported which are in disagreement with the $n/3$ law for $n > 3$.

In fact, in neither the (1941a, b) papers of Kolmogorov, referenced by Academician U. Frisch, nor the other papers of Kolmogorov was an $n/3$ law proposed. Moreover, the applicability of the vanishing-viscosity principle to higher moments seems to be doubtful. The scaling laws obtained by the above-mentioned experimentalists can be explained by the fact that they worked on small set-ups, and so the universality of the results cannot be assumed.

The spectral (wave vector) representation of the correlation moments is also of interest; it is especially popular in the physics literature. Therefore it is appropriate to give here the spectral form of the basic relation (12.37). Denote by $E(k)\,dk$ the energy of vortices whose wave numbers (the inverse length scales of vortices) lie in the range k, $k + dk$. Then $E(k)$ should depend on the governing parameters k (instead of r), ε, ν and Re. Advanced similarity analysis gives, in agreement with the existence of incomplete similarity in the parameter $k\lambda$, the relation

$$E(k) = C(Re)\varepsilon^{2/3}\,k^{-5/3}(k\lambda)^{\gamma/\ln Re} . \qquad (12.51)$$

Experiments performed at large but still moderate Reynolds numbers show that, up to accuracy of the data, the exponent is close to $-5/3$. This means that the contribution of the term $\gamma/\ln Re$ in the exponent can be neglected to a first approximation. However, the Re-dependence of C is noticeable, so at large but moderate Reynolds numbers the following form of the spectral law can be proposed (the 5/3 law):

$$E(k) = (C_0 + C_1/\ln Re)\varepsilon^{2/3}\,k^{-5/3} , \qquad (12.52)$$

which is analogous to the formula for D_{11} (the 2/3 law):[3]

$$D_{11} = (A_0 + A_1/\ln Re)(\varepsilon r)^{2/3} . \qquad (12.53)$$

A historical comment It is generally recognized that it was A. N. Kolmogorov who shaped the modern theory of turbulence. It is clear that the next step of similar strength will require a genius of comparable power and, what is also very important, comparable especially favorable circumstances. Here it is appropriate to remember these circumstances and two outstanding people who played very significant roles in the phenomenon of Kolmogorov's discoveries in turbulence.

---

[3] It is important to emphasize that in the literature, starting from the original articles by A. N. Kolmogorov and A. M. Obukhov, the 2/3 and $-5/3$ laws were derived using the assumption (similar to that of von Kármán) that in the inertia interval the correlation moments are viscosity independent. The 2/3 and $-5/3$ laws have been well confirmed by experiments. However, the numerical factor in front of $(\varepsilon r)^{2/3}$ or $(k\varepsilon)^{-5/3}$ obtained in experiments was, unexpectedly found not to be a universal constant. This shows that, at very large Re, the correction to 2/3 (or $-5/3$) is small in comparison with unity, and so there is no contradiction with our assumption of incomplete similarity.

The first was Otto Yulievich Schmidt. This man seemed to have been transferred to the first half of the twentieth century from the Renaissance epoch. In the late 1920s and early 1930s he had already gained remarkable achievements in several different fields. In 1932 he was Professor of Algebra at Moscow State University, Director of the State Publishing House, the only one in the country, Editor-in-Chief of the *Soviet Encyclopedia*, Deputy People's Commissar (Minister) of Finances, had other important duties, and was successful everywhere.

Unexpectedly for everybody including himself he was then appointed... the Chief of the Directorate of the Northern Sea Path! The situation regarding northern navigation was critically important for the country. The highest authority recognized it and came to the conclusion that Schmidt, if properly motivated, could achieve success. Indeed, in two to three years Schmidt transformed miraculously the decaying field of northern navigation into a flourishing and attractive part of Soviet life. Very importantly, the unpopular profession of polar explorer quickly became prestigious and fashionable. Two bright events made the name of Schmidt especially popular, even legendary: the polar expedition on the ship "Chelyuskin" (1934) and the expedition to the North Pole (1937–1938) which he planned and personally guided. In the spring of 1938, when the ice floe on which the participants in the North Pole expedition worked started to melt and break, Schmidt headed the expedition to save the team and the scientific results of the expedition, landed on the ice floe and was one of the last to leave it. The popularity of Schmidt in the country and in the whole world became an incomparable symbol of Soviet prestige, as would the "Sputnik" voyage into space 20 years later.

After that Schmidt obtained a new position: he was appointed the plenipotentiary first Vice-President of the Soviet Academy of Sciences (the President was ill). The first thing that Schmidt recognized in his new position was the need for a strong supportive team of genuinely outstanding scientists – members of the Academy. The first person whom he addressed was A. N. Kolmogorov, his colleague in the Department of Mathematics of Moscow State University, Professor of Probability Theory. Schmidt offered him the position of a full Member of the Academy and, moreover, that of the Head (Secretary) of the Division of Physical and Mathematical Sciences of the Academy. (At that time Kolmogorov was not even a Corresponding Member). Schmidt asked one favor of Kolmogorov: to join the staff of the Institute of Theoretical Geophysics, which Schmidt planned to organize, and to start a laboratory there. The idea was that Kolmogorov would bring to this laboratory some of his disciples and, after a time, leave the laboratory in the charge of one of them.

And so it happened that, following Schmidt's proposal, suggested, of course, in a very delicate form, as one possibility, the basic subject of the laboratory became turbulence. (What a fantastic vision of Schmidt: Kolmogorov was known at that time as the rising star of pure mathematics!) The first and leading Kolmogorov disciple who entered the laboratory was A. M. Obukhov, followed later by A. M. Yaglom, A. S. Monin, G. S. Golitsyn, V. I. Tartarsky, V. I. Klyantskin, F. V. Dolzhansky, E. B. Gledser, A. S. Gurvich, B. M. Bubnov, more recently O. G. Chkhetiani and other remarkable scientists. The first seminar delivered by A. N. Kolmogorov devoted to the new subject was in September 1939. The laboratory developed and transformed itself into the flourishing A. M. Obukhov Institute of Atmospheric Physics of the Russian Academy of Sciences.

The second person who played a very important role in the Kolmogorov turbulence phenomenon was George Keith Batchelor, one of the future leaders of the post-war fluid mechanics. It happened that he came to Cambridge from his home country Australia at the very end of World War II (April 1945) in a convoy consisting of 80 ships. His goal was to work at the Cavendish Laboratory with G. I. Taylor on turbulence. When Batchelor met Taylor, he realized that Taylor was no longer interested in turbulence. However, turbulence continued to attract the attention of Batchelor, and he started to search in the Library of Cambridge University for research sources concerning this subject. Batchelor found the volumes of *The Proceedings (Doklady) of the Soviet Academy of Sciences*, by a miracle delivered to Cambridge during the war, and found there the first papers of Kolmogorov. He understood immediately what a treasure was in his hands!

Batchelor's further activity was unbelievably fast and productive. He prepared and delivered a lecture about Kolmogorov–Obukhov theory at the International Congress on Applied Mechanics in the following year, 1946. Moreover, he performed a careful analysis of the text of the short notes where Kolmogorov–Obukhov theory was presented, thoroughly clarified difficult places and prepared a detailed explanation of the theory in a comprehensive article (Batchelor, 1947). This article played a decisive role in the propagation of Kolmogorov–Obukhov theory. In fact, during the 20 years preceding the appearance of the fundamental monograph of Monin and Yaglom, Batchelor's article, as well as the first (1944) edition of Landau and Lifshitz's book *Mechanics of Continuous Media*, available only in Russian, remained the only source where a detailed presentation of the theory of the local structure of turbulent flows at very large Reynolds numbers could be found.

Bibliography and References

The reading list consists of two parts. The first part comprises monographs of fundamental importance that are appropriate for more detailed study and even for learning about the various different viewpoints.

The second part consists of articles and books referred to in the text.

Fundamental monographs

Arnold, V. I. (1978). *Mathematical Methods of Classical Mechanics*. Springer-Verlag.

Batchelor, G. K. (2002). *An Introduction to Fluid Dynamics*. Cambridge University Press.

Bažant, Z. P. (2002). *Scaling of Structural Strength*. Hermes-Penton Science.

Bridgman, P. W. (1931). *Dimensional Analysis*. Yale University Press.

Broberg, K. B. (1999). *Cracks and Fracture*. Academic Press.

Carpinteri, A. (ed.) (1996). *Size-Scale Effects in the Failure Mechanisms of Materials and Structures*. E. & F. N. Spon.

Chernyi, G. G. (1961). *Introduction to Hypersonic Flows*. Academic Press.

Chorin, A. J. (1994). *Vorticity and Turbulence*. Springer-Verlag.

Chorin, A. J., and Marsden, J. E. (1992). *A Mathematical Introduction to Fluid Mechanics*, 3rd edn. Springer-Verlag.

Feynman, R. (2006). *The Feynman Lectures on Physics*, definitive edn. Addison-Wesley.

Friedrichs, K. O. (1966). *Special Topics in Fluid Dynamics*. Gordon and Breach.

Frisch, U. (1996). *Turbulence. The Legacy of A. N. Kolmogorov*. Cambridge University Press.

Galilei, Galileo (1638). *Discorsi e Dimonstrazioni Matematiche Intorno á Duo Nuove Scienze*. Elsevier, Leida. Also: *Dialogues Concerning Two New Sciences*. Easton Press (1999).

Germain, P. (1986). *Mécanique*. École Polytechnique. Ellipses.

Goldenfeld, N. D. (1992). *Lectures on Phase Transitions and the Renormalization Group*. Perseus Publishing.

Hayes, W. D., and Probstein, R. F. (2004). *Hypersonic Inviscid Flow*. Dover Publications.

Lamb, H. (1932). *Hydrodynamics*, 6th edn. Cambridge University Press.

Landau, L. D., and Lifshitz, E. M. (1986). *Theory of Elasticity*, 3rd edn. Elsevier.

Landau, L. D., and Lifshitz, E. M. (1987). *Fluid Mechanics*, 2nd edn. Elsevier.

Lighthill, M. J. (1986). *An Informal Introduction to Theoretical Fluid Mechanics*. Oxford University Press.

Love, A. E. H. (1944). *A Treatise on the Mathematical Theory of Elasticity*, 4th edn. MacMillan.

Mandelbrot, B. (1975). *Les Objets Fractals: Forme, Hasard et Dimension*. Flammarion, Paris.

Mandelbrot, B. (1977). *Fractals, Form, Chance and Dimension*. W. H. Freeman & Co.

Marsden, J. E., and Hughes, T. J. R. (1983). *The Mathematical Foundations of Elasticity*. Prentice Hall.

Monin, A. S., and Yaglom, A. M. (1975). *Statistical Fluid Mechanics, Mechanics of Turbulence, Vol. II*. MIT Press.

Monin, A. S., and Yaglom, A. M. (1971). *Statistical Fluid Mechanics, Mechanics of Turbulence, Vol. I*. MIT Press.

Muskhelishvili, N. I. (1963). *Some Basic Problems of Mathematical Theory of Elasticity*, 2nd English edn. P. Nordhoft.

Oswatisch, K. (1956). *Gas Dynamics*. Academic Press.

Prandtl, L., and Tietjens, O. (1931). *Hydro und Aeromechanik, Vol. 2*. Springer-Verlag.

Schlichting, H. (1968). *Boundary Layer Theory*, 6th edn. McGraw-Hill.

Suresh, S. (1998). *Fatigue of Materials*. Cambridge University Press.

Timoshenko, S. P. (1953). *History of Strength of Materials*. McGraw-Hill.

Timoshenko, S. P., and Goodier, J. N. (1970). *Theory of Elasticity*. McGraw-Hill.

Todhunter, I., and Pearson, K. (1886). *A History of the Theory of Elasticity and of the Strength of Materials, Vol. I*. Cambridge University Press.

Van Dyke, M. (1982). *An Album of Fluid Motion*, 10th edn. Parabolic Press.

von Kármán, Th. (1957). *Aerodynamics*. Cornell University Press.

Whitham, G. B. (1974). *Linear and Non-linear Waves*. J. Wiley & Sons.

Zeldovich, Ya. B., and Raizer, Yu. P. (2002). *Physics of Shock Waves and High Temperature Hydrodynamic Phenomena*. Dover Publications.

References

Adamsky, V. B. (1956). Integration of the system of self-similar equations in the problem of an impulsive load on a cold gas. *Akust. Zh.* **2**, 3–9.

Atiyah, M. F., Bott, R., and Gårding, L. (1970). Lacunae for hyperbolic differential operators with constant coefficients. *Acta Math.* **124**, 109–189.

Barenblatt, G. I. (1956). On certain problems of the theory of elasticity arising in the study of the mechanism of the hydraulic fracture of oil-bearing strata. *J. Appl. Math. Mech. (PMM)* **20** (4), 475–486.

Barenblatt, G. I. (1959). On the equilibrium cracks formed in brittle fracture. *J. Appl. Math. Mech. (PMM)* **23** (3), 434–444; (4) 706–721; (5) 893–900.

Barenblatt, G. I. (1962). The mathematical theory of equilibrium cracks in brittle fracture. In *Advances in Applied Mechanics*, H. L. Dryden and Th. von Kármán (eds.), Vol. VII, pp. 55–129.

Barenblatt, G. I. (1964). On some general concepts of the mathematical theory of brittle fracture. *J. Appl. Math. Mech. (PMM)* **28** (4), 778–792.

Barenblatt, G. I. (1996). *Scaling, Self-Similarity and Intermediate Asymptotics*. Cambridge University Press.

Barenblatt, G. I. (2003). *Scaling*. Cambridge University Press.

Barenblatt, G. I. (2008). A mathematical model of turbulent drag reduction by high-molecular-weight polymeric additions in a shear flow. *Phys. Fluids* **20**, 091 702.

Barenblatt, G. I., and Botvina, L. R. (1981). Incomplete similarity of fatigue in a linear range of crack growth. *Fatigue Eng. Mater. Struct.* **3**, 193–212.

Barenblatt, G. I., and Chernyi, G. G. (1963). On the moment relations on the discontinuity surfaces in dissipative media. *J. Appl. Math. Mech. (PMM)* **27** (5), 1205–1218.

Barenblatt, G. I., Chorin, A. J., and Prostokishin, V. M. (1997). Scaling laws in fully developed turbulent pipe flow. *Appl. Mech. Rev.* **50**, 413–429.

Barenblatt, G. I., Chorin, A. J., and Prostokishin, V. M. (2000). Self-similar intermediate structures in turbulent boundary layers at large Reynolds numbers. *J. Fluid Mech.* **410**, 263–283.

Barenblatt, G. I., Chorin, A. J., and Prostokishin, V. M. (2002). A model of turbulent boundary layer with non-zero pressure gradient. *Proc. US Nat. Acad. Sci.* **99**, 5572–5576.

Barenblatt, G. I., Entov, V. M., and Salganik, R. L. (1966). Kinetics of crack extension. *Eng. J. Mech. Solids* **1** (5), 53–69; **1** (6), 49–51.

Barenblatt, G. I., and Goldenfeld, N. D. (1995). Does fully developed turbulence exist? Reynolds number independence versus asymptotic covariance. *Phys. Fluids* **7** (12), 3078–3082.

Barenblatt, G. I., and Monteiro, P. J. M. (2010). Scaling laws in nanomechanics. *Physical Mesomech.* **13** (5–6), 245–248.

Batchelor, G. K. (1947). Kolmogoroff's theory of locally isotropic turbulence. *Proc. Camb. Phil. Soc.* **43** (4), 533–559.

Batchelor, G. K. (1996). *The Life and Legacy of G. I. Taylor.* Cambridge University Press.

Bažant, Z. P. (2002). *Scaling of Structural Strength.* Hermes-Penton Science.

Benbow, J. J. (1960). Cone cracks in fused silica. *Proc. Phys. Soc.* **B75**, 697–699.

Blasius, H. (1908). Grenzschichten in Flüssigkeit mit kleiner Reibung. *Z. Math. Phys.* **56**, 1–37.

Bok, B. J. (1972). The birth of stars. *Scientific American* **227** (2), 48–65.

Botvina, L. R. (1989). *The Kinetics of Fracture of Structural Materials,* Nauka.

Boussinesq, J. (1877). Essai sur la théorie des eaux courants. *Mémoirs Présentés par Divers Savants a l'Académie des Sciences, Paris* **23** (1), 1–680.

Carpinteri, A. E. (1996). Strength and toughness in disordered materials: complete and incomplete similarity. In *Size Scale Effects in the Failure Mechanisms of Materials and Structures,* A. E. Carpinteri (ed.), pp. 3–26. E. & F. N. Spon.

Castaing, B., Gagne, Y., and Hopfinger, E. (1990). Velocity probability dencity functions of high Reynolds number turbulence. *Physica D* **46**, 177–220.

Cauchy, A. L. (1828). Sur les équations qui experiment les conditions d'équilibre ou les lois du movement des fluides. In *Exercises de Mathématiques,* Bure, Paris. Also in *Oevres Complètes d'Augustin Cauchy,* II Série, Tom VIII, pp. 158–179. Gauthier-Villars (1890).

Chaplyguine, S. A. (1910). On the pressure of the plane-parallel flow on the blocking bodies (on the theory of aeroplanes). *Mat. Sbornik* **28**, Moscow.

Chernyi, G. G. (1961). *Introduction to Hypersonic Flows.* Academic Press.

Chorin, A. J. (1977). Theories of turbulence. In *Lecture Notes in Mathematics,* Vol. 615, pp. 36–47. Springer-Verlag.

Chorin, A. J. (1994). *Vorticity and Turbulence.* Springer-Verlag.

Chorin, A. J. (1998). New perspectives in turbulence. *Quart. J. Appl. Math.* **XIV** (4), 767–785.

Cottrell, A. H. (1967). The nature of metals. *Scientific American* **217** (3), 90–100.

Dugdale, D. S. (1960). Yielding of steel sheets containing slits. *J. Mech. Phys. Solids* **8**, 100–104.

Ekman, V. W. (1910). On the change from steady to turbulent motion of liquids. *Ark. Mat. Astronom. Fys.* **6** (12).

Erm, L. P., and Joubert, P. N. (1991). Low Reynolds-number turbulent boundary layers. *J. Fluid Mech.* **230**, 1–44.

Fernholz, H. H., and Finley, P. J. (1996). The incompressible zero-pressure gradient turbulent boundary layer: an assessment of the data. *Progr. Aero. Sci.* **32**, 245–311.

Gilman, J. J. (1967). The nature of ceramics. *Scientific American* **217** (3), 113–124.

Goldenfeld, N. D. (1992). *Lectures on Phase Transitions and the Renormalization Group*. Perseus Publishing.

Goodier, J. N. (1968). Mathematical theory of equilibrium cracks. In *Fracture. An Advanced Treatise, Vol. II*, H. Liebowitz (ed.), pp. 1–66.

Griffith, A. A. (1920). The phenomenon of rupture and flow in solids. *Phil. Trans. Roy. Soc. London* **A221**, 163–198.

Griffith, A. A. (1924). The theory of rupture. In *Proc. 1st Int. Congress on Applied Mathematics*, Delft, pp. 55–63.

Harmon, L. D. (1973). Recognition of faces. *Scientific American* **229** (5), 70–82.

Hayden, H. W., Gibson, R. C., and Brophy, J. H. (1969). Superplastic steels. *Scientific American* **220** (3), 28–35.

Heisenberg, W. (1948). On the theory of statistical and isotropic turbulence. *Proc. Roy. Soc. London* **A195**, 402–406.

Hooke, R. (1678). *De Potentia Restitutiva*. London.

Hunt, J. C. R., Phillips, O. M., and Williams, D. (eds.) (1991). Turbulence and stochastic processes: Kolmogorov's ideas 50 years on. *Proc. Roy. Soc. London* **434**.

Irwin, G. R. (1948). Fracture dynamics. In *Fracturing of Metals*, pp. 147–166. ASM, Cleveland, Ohio.

Irwin, G. R. (1957). Analysis of stresses and strains near the end of a crack traversing a plate. *J. Appl. Mech.* **24**, 361–364.

Irwin, G. R. (1960). Plastic zone near a crack and fracture toughness. In *Mechanical and Metallurgical Behaviour of Sheet Materials*, Proc. Seventh Sagamore Conf.

Izakson, A. (1937). Formula for the velocity distribution near a wall. *Zh. Exper. Teor. Fiz.* **7** (7), 919–924.

Joseph, D., Funada, T., and Wang, J. (2008). *Potential Flows of Viscous and Viscoelastic Fluids*. Cambridge University Press.

Joukovsky, N. E. (1906). On attached vortices. *Proc. Section of the Physical Sciences XIII*, no. 2. Typography of Moscow University.

Joukovsky, N. E. (1910). Ueber die Konturen der Tragflächen der Drachenfliege. *Z. für Flugtechnik and Motorluftschiffahrt* **I** (22), 281–284.

Kelly, A. (1967). The nature of composite materials. *Scientific American* **217** (3), 160–179.

Kline, S. J., Reynolds, W. C., Schraub, W. A., and Rundstadler, P. W. (1967). The structure of turbulent boundary layers. *J. Fluid Mech.* **30** (1), 741–773.

Kolmogorov, A. N. (1941a). Local structure of turbulence in an incompressible fluid at very high Reynolds numbers. *Dokl. Akad. Nauk SSSR* **30** (4), 299–303. Also in *Selected Works of A. N. Kolmogorov, Vol. I*, pp. 312–318. Kluwer (1991).

Kolmogorov, A. N. (1941b). Energy dissipation in locally isotropic turbulence. *Dokl. Akad. Nauk SSSR* **31** (1), 19–21. Also in *Selected Works of A. N. Kolmogorov, Vol. I*, pp. 324–327. Kluwer (1991).

Kolmogorov, A. M. (1942). The equations of turbulent motion of incompressible fluids. *Izvestiya, Akad. Nauk SSSR, Ser. Fiz.* **6** (1–2), 56–58. Also in *Selected Works of A. N. Kolmogorov, Vol. I*, pp. 328–330. Kluwer (1991).

Kolosov, G. V. (1909). On the application of complex functions theory to a plane problem of the mathematical theory of elasticity. Thesis, Yuriev (Dorpat) University.

Krogstad, P. A., and Antonia, P. A. (1999). Surface roughness effects in turbulent boundary layers. *Exp. Fluids* **27**, 450–460.

Kudin, A. M., Barenblatt, G. I., Kalashnikov, V. N., Vlasov, S. A., and Belokon', V. S. (1973). Destruction of metallic obstacles by a jet of dilute polymer solution. *Nature (London), Phys. Sci.* **245**, 95.

Kutta, W. M. (1902). Auftriebskräfte in strömenden Flüssigkeiten. In *Illustrierte Aeronautische Mitteilungen*, München, p. 133.

Kutta, W. M. (1910). Uber eine mit den Grundlagen des Flugproblems in Bezielung stellende zweidimensionale Strömung. *Sitzungberichte der Königlich Bayerischen Akademie der Wissenschaften. Mathematisch-Physikalische Klasse* **2**, 1–58.

Lax, P. (1968). Integrals of nonlinear equations of evolution and solitary waves. *Comm. Pure Appl. Math.* **21**, 467–490.

Leonov, M. Ya., and Panasyuk, V. V. (1959). Development of the finest cracks in a solid. *Prikladna Mech.* **5**, 391–401.

Lighthill, M. J. (1970). Turbulence. In *Osborne Reynolds and Engineering Science Today*, D. M. McDowell and J. D. Jackson (eds.), Manchester University Press. Also in *Collected Papers of Sir James Lighthill, Vol. II*, pp. 83–146. Oxford University Press (1997).

Mark, H. F. (1967). The nature of the polymeric materials. *Scientific American*, **217** (3), 149–156.

Marušić, I., and Perry, A. E. (1995). A wall-wake model for the turbulence structure of boundary layers. Part 2. Further experimental support. *J. Fluid. Mech.* **298**, 389–407.

Millikan, C. B. (1939). A critical discussion of turbulence in channels and circular pipes. In *Proc. 5th Int. Congress in Applied Mechanics*, Cambridge, MA, pp. 386–392.

Monin, A. S., and Yaglom, A. M. (1971). *Statistical Fluid Mechanics, Mechanics of Turbulence, Vol. I.* MIT Press.

Mott, Sir Nevill (1967). The solid state. *Scientific American* **217** (3), 80–89.

Navier, C. L. M. H. (1822). Mémoire sur les lois du mouvement des fluides. *Mémoires de l'Académie Royale des Sciences de l'Institut de France* **6**, 389–449.

Navier, C. L. M. H. (1827). Mémoire sur les lois de l'equilibre et du mouvement des corps solides élastiques. *Mémoires de l'Académie Royale des Sciences de l'Institut de France* **7**, 375–393.

Nigmatulin, R. I. (1965). A plane strong explosion on a boundary of two ideal, calorically perfect gases. *Bulletin MGU, Ser. Matem. Mech.* **1**, 83–87.

Nikuradze, J. (1932). Gesetzmässigkeiten der turbulenten Strömung in glatten Röhren. *VDI Forschungsheft*, no. 356.

Obukhov, A. M. (1941a). Spectral energy distribution in a turbulent flow. *Dokl. Akad. Nauk SSSR* **32** (1), 22–24.

Obukhov, A. M. (1941b). Spectral energy distribution in a turbulent flow. *Izvestiya Akad. Nauk SSSR, Ser. Geogr. Geofiz.* **5** (4–5), 453–466.

Onsager, L. (1945). The distribution of energy in turbulence. *Phys. Rev.* **68** (11–12), 286.

Orowan, E. O. (1950). Fundamentals of brittle behavior of metals. In *Fatigue and Fracture of Metals*, W. M. Murray (ed.), pp. 139–167. Wiley.

Oswatisch, K. (1956). *Gas Dynamics.* Academic Press.

Panasyuk, V. V. (1968). *Limiting Equilibrium of Brittle Bodies with Cracks.* Naukova Dumka, Kiev.

Panton, R. I. (2002). Evaluation of the Barenblatt–Chorin–Prostokishin power law for boundary layers. *Phys. Fluids* **14** (5), 1806–1808.

Paris, P. C., and Erdogan, F. (1963). A critical analysis of crack propagation laws. *J. Basic Eng. Trans. ASME, Ser. D* **85**, 528–534.

Perry, A. E., Hafer, S., and Chong, M. S. (2001). A possible reinterpretation of the Princeton superpipe data. *J. Fluid Mech.* **439**, 395–401.

Petrovsky, I. G. (1966). *Ordinary Differential Equations.* Prentice Hall.

Petrovsky, I. G. (1967). *Partial Differential Equations.* W. G. Saunders.

Petrovsky, I. G. (1971). *Lectures on the Theory of Integral Equations.* Mir Publishers.

Poisson, S. D. (1829). Sur les équations général de l'equilibre et du mouvement des corps solides élastiques et des fluides. *J. de l'École Royale Polytechnique* **13**, 1–174.

Prandtl, L. (1905). Uber Flüssigkeits Bewegung bei sehr kleiner Reibung. In *Verhandlungen des III Int. Math. Kongress (Heidelberg, 1904)*, pp. 484–491.

Teubner, Leipzig. Also in *Ludwig Prandtl: Gesammelte Abhandlungen*, W. Tollmien, H. Schlichting, and H. Görtler (eds.), pp. 575–584. Springer-Verlag.

Prandtl, L. (1932). Zur turbulenten Strömung in Röhren and längs Platten. *Ergebn. Aerodyn. Versuchsanstalt, Göttingen* **B4**, 18–29.

Prandtl, L. (1945). Uber ein neues Formalsystem für die ausgebildete Turbulenz. *Nacht. Akad. Wiss. Göttingen, Math.-Phys. Klasse*, 6–18.

Prandtl, L., and Tietjens, O. (1931). *Hydro und Aeromechanik, Vol. 2*. Springer-Verlag.

Praskovsky, A. S., and Oncley, S. (1994). Measurement of Kolmogorov constant and intermittency exponent at very high Reynolds numbers. *Phys. Fluids* **6** (9), 2886–2889.

Rayleigh, Lord J. W. (1877). On the irregular flight of a tennis ball. *Messenger of Mathematics* **VII**, 14–16. Also in *Scientific Papers, Vol. 1*, pp. 344–346. Cambridge University Press (1899).

Reynolds, O. (1883). An experimental investigation of the circumstances which determine whether the motion of water shall be direct or sinuous and the law of resistance in a parallel channel. *Phil. Trans. Roy. Soc. London* **174**, 935–982.

Reynolds, O. (1894). On the dynamical theory of incompressible viscous fluids and the determination of the criterion. *Phil. Trans. Roy. Soc. London* **186**, 123–161.

Richardson, L. F. (1922). *Weather Prediction by Numerical Process*. Cambridge University Press.

Ritchie, R. O. (2005). Incomplete self-similarity and fatigue crack growth. *Int. J. Fracture* **132**, 97–203.

Ritchie, R. O., and Knott, J. F. (1974). Micro cleavage cracking during fatigue crack propagation in low strength steel. *Mat. Sci. Engng.* **14**, 7–14.

Roesler, F. (1956). Brittle fracture near equilibrium. *Proc. Phys. Soc.* **B69**, 981–992.

Russell, J. S. (1844). *Report on Waves*. In *Reports of the* XIV *Meeting of the British Association for the Advancement of Science*. J. Murray.

Saint-Venant, Barré de, A. J. C. (1843). Note á joindre au mémoire sur la dynamique des fluides presenté le 14 avril 1834. *Comptes Rendus de l'Académie des Sciences* **17** (22), 1240–1243.

Schiller, L. (1922). Untersuchungen über laminare und turbulente Strömung. *Forschungarbeiten Ing.-Wesen H.* **248**; *ZAMM* **2**, 96–106.

Schlichting, H. (1968). *Boundary Layer Theory*, 6th edn. McGraw-Hill.

Sedov, L. I. (1946). Propagation of strong shock waves. *J. Appl. Math. Mech. (PMM)* **10**, 241–250. (Pergamon Translation no. 1223).

Sedov, L. I. (1959). *Similarity and Dimensional Methods in Mechanics*. Academic Press.

Spurk, J. H., and Aksel, N. (2008). *Fluid Mechanics*. Springer-Verlag.

Stokes, G. G. (1845). On the theories of the internal friction of fluids in motion, and of the equilibrium and motion of elastic solids. *Trans. Camb. Phil. Soc.* **VII** (I), 287–319.

Taylor, G. I. (1941). The formation of a blast wave by a very intense explosion. Report RC–210, 27 June 1941, Civil Defense Research Committee.

Taylor, G. I. (1950). The formation of a blast wave by a very intense explosion. II. The atomic explosion of 1945. *Proc. Roy. Soc. London* **A201**, 175–186.

Teixeira, R. E., Babcock, H. P., Shaqfeh, E. S. G., and Chu, S. (2005). Shear thinning and tumbling dynamics of single polymers in the flow gradient plane. *Macromolecules* **38**, 581.

Teixeira, R. E., Dambel, A. K., Richter, D. H., Shaqfeh, E. S. G., and Chu, S. (2007). The individualistic dynamics of entangled DNA in solution. *Macromolecules* **40**, 2461.

Töpfer, C. (1912). Anmerkungen zu dem Aufsatz von H. Blasius "Grenzschichten in Flüssigkeit mit kleiner Reibung". *Z. Math. Phys.* **60**, 397.

Trefil, J. (1999). *Other Worlds. Images of the Cosmos from Earth and Space*. National Geographic.

Van den Booghart, A. (1966). Crazing and characterization of brittle fracture in polymers. In *Proc. Conf. Physical Basis of Yield and Fracture*. Oxford University Press.

Vlasov, I. O., Derzhavina, A. I., and Ryzhov, O. S. (1974). On an explosion on the boundary of two media. *Comput. Math. and Math. Phys.* **14** (6), 1544–1552.

von Kármán, Th. (1930). Mechanische Ähnlichkeit und Turbulenz. In *Proc. III Int. Congr. Applied Mechanics*, C. W. Oseen and W. Weibull (eds.), Vol. 1, pp. 81–93. AB Sveriges Litografska Truckenier.

von Kármán, Th., and Howarth, L. (1938). On the statistical theory of isotropic turbulence. *Proc. Roy. Soc. London* **A164** (917), 192–215.

von Koch, H. (1904). Sur une courbe continue sans tangente obtenue par une construction géometrique élémentaire. *Arkiv Mat. Astron. Fys.* **2**, 681–704.

von Mises, R. (1941). Some remarks on the laws of turbulent motion in tubes. In *Th. von Kármán Anniversary Volume*, pp. 317–327. California Institute of Technology Press.

von Neumann, J. (1941). The point source solution. National Defense Research Committee, Div. B, Report AM–9, 30 June 1941.

von Weizsäcker, C. F. (1948). Der Spektrum der Turbulenz bei grossen Reynolds'schen Zahlen. *Z. Physik* **124** (7–12), 614–627.

von Weizsäcker, C. F. (1954). Genäherte Darstellung starker instationäzer Stosswellen durch Homologie-Lösungen. *Z. Naturforschung* **9A**, 269–275.

Willis, J. R. (1967). A comparison of the fracture criteria of Griffith and Barenblatt. *J. Mech. Phys. Solid* **15**, 151–162.

Wu, X., and Moin, P. (2009). Direct numerical simulation of turbulence in a nominally zero-pressure gradient flat plate boundary layer. *J. Fluid Mech.* **630**, 5–41.

Yaglom, A. M. (1993). Similarity laws for wall turbulent flows: their limitation and generalizations. In *Conf. on New Approaches and Concepts in Turbulence*, Monte Verita, Th. Dracos and A. Tsinober (eds.), pp. 7–27. Birkhäuser-Verlag.

Yaglom, A. M. (2000). The century achievements and unsolved problems. In *New Trends in Turbulence*, M. Lesieur, A. Yaglom and F. David (eds.), pp. 3–52. Springer-Verlag.

Zagarola, M. V. (1996). Mean flow scaling in turbulent pipe flow. Ph.D. thesis, Princeton University.

Zagarola, M. V., Smits, A. J., Orszag, S. A., and Yakhot, V. (1996). Experiments in high Reynolds number turbulent pipe flow. AIAA paper 95–0654, Reno, NV.

Zeldovich, Ya. B. (1956). The motion of gas under the action of a short term pressure shock. *Akust. Zh.* **2** (1), 28–30. (*Soviet Phys. Acoustics* **2**, 25–35.)

Zeldovich, Ya. B., and Raizer, Yu. P. (2002). *Physics of Shock Waves and High Temperature Hydrodynamic Phenomena*. Dover Publications.

Zheltov, Yu. P., and Christianovich, S. A. (1955). On the hydraulic fracture of oil strata. *Izvestiya, USSR Acad. Sci., Technical Sci.* **5**, 3–41.

Zhukov, A. I., and Kazhdan, Ia. M. (1956). Motion of a gas due to the effect of a brief impulse. *Soviet Phys. Acoustics* **2** (4), 375–381.

Index

Printed in the United States
By Bookmasters